ELEMENTS OF THE
TOPOLOGY OF PLANE SETS
OF POINTS

T0292052

ELEMENTS OF THE
TOPOLOGY OF PLANE SETS
OF POINTS

by

M. H. A. NEWMAN, M.A., F.R.S.

Fielden Professor of Mathematics
in the University of Manchester

CAMBRIDGE
AT THE UNIVERSITY PRESS
1964

CAMBRIDGE UNIVERSITY PRESS
Cambridge, New York, Melbourne, Madrid, Cape Town, Singapore, São Paulo, Delhi

Cambridge University Press
The Edinburgh Building, Cambridge CB2 8RU, UK

Published in the United States of America by Cambridge University Press, New York

www.cambridge.org
Information on this title: www.cambridge.org/9780521058117

First edition 1939
Second edition, re-set 1951
Reprinted 1954, 1961, 1964
Re-issued in this digitally printed version 2008

A catalogue record for this publication is available from the British Library

ISBN 978-0-521-05811-7 hardback
ISBN 978-0-521-09186-2 paperback

CONTENTS

PREFACE TO THE SECOND EDITION

Although substantial changes have been made throughout the book, its aim is as before, to provide an elementary introduction to the ideas and methods of topology by the detailed study of certain topics, with special attention to the parts needed in the theory of functions.

The contents of the former Part II have been rearranged, the "homology" theory of simply- and multiply-connected domains occupying Chapter VI, and the "homotopy" theory (paths and their deformations) Chapter VII. The treatment of homology on a grating has been modified so as to make clear the distinction between "chains" and their point-set loci. The sections on boundary elements of domains, and on the connectivity of certain kinds of closed sets, have been omitted, and their place taken by a section on the orientation of plane curves. A number of topics, such as the properties of vector spaces of infinite dimension, the Jacobian theorems on implicit functions, and the Cauchy integral theorem (with one or more boundary curves), are treated as examples.

Notations are, on the whole, unchanged, except that in the algebra of sets the symbols $\cup$ and $\cap$ have replaced $+$ and $\cdot$, in accordance with current usage.

I am grateful to many correspondents, and particularly to M. Richardson, F. Smithies, S. Wylie and D. G. Northcott, for pointing out errors in the first edition, or suggesting improvements; to my colleagues A. H. Stone, P. Hilton, W. Ledermann and G. E. H. Reuter for reading the revised text in manuscript and proof, and making many valuable suggestions; and to the Cambridge University Press for the great care that they have taken with the printing.

M. H. A. NEWMAN

MANCHESTER 1951

Chapter I

SETS*

§ 1. THE CALCULUS OF SETS

1. The object of the calculus described in § 1 is a practical one—to shew how complicated properties of sets may be deduced by formal rules from a small number of properties which are sufficiently simple to be self-evident. The propositions accepted as self-evident (lettered A to F) are not intended as a system of axioms—a much smaller number would suffice for that purpose—but as a convenient body of "standard forms" for use in symbolical work.

2. A set (or class or aggregate) is to be thought of not as a heap of things specified by enumerating its members one after another, but as something determined by a *property*, which can be used to test the claim of any object to be a member of the set. Thus the set of even integers is determined by the property of being twice some other integer, the algebraic numbers by the property of satisfying a polynomial equation with integral coefficients. Two properties determine the same set if they are "formally equivalent", i.e. if no object has one property without having the other.

The symbol $x \in A$ means "x is a member (or element) of A". The set which has only the single member a is denoted by (a). Thus $x \in (a)$ means simply $x = a$.

3. The symbol $A \subseteq B$ (or $B \supseteq A$) means that, for every x,

$$x \in A \quad \text{implies} \quad x \in B.$$

It is usually read "A is contained in B", or "A is a subset of B", but, as the form of the symbol suggests, identity is not excluded.

---

* The subjects dealt with in this preliminary chapter are (1) the calculus or algebra of sets, and (2) the distinction between enumerable sets and others. Readers who are familiar with these matters should omit this chapter, but should note the definitions here adopted for the symbols $A \subseteq B$ and $A \subset B$, pp. 1, 2; $B-A$, p. 5; and $\mathscr{C}A$, p. 5.

The symbol $A \subset B$ (rarely used in this book) means "$A \subseteq B$ but $A \neq B$", and is read "A is a proper subset of B".[1]*

A 1. $A \subseteq A$.

2. *If $A \subseteq B$ and $B \subseteq C$ then $A \subseteq C$.*

3. *$A = B$ if, and only if, $A \subseteq B$ and $B \subseteq A$.*

These properties of sets must be accepted as self-evident. (**A** 3 may, if preferred, be regarded as a definition of equality between sets.)

4. The *join* or *union*, $A \cup B$, of the sets A and B is the set of all members of either set. Thus "$x \in A \cup B$" means "$x \in A$ or

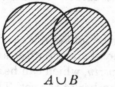

$A \cup B$
Fig. 1

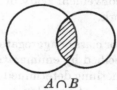

$A \cap B$
Fig. 2

$x \in B$ (or both)". For example, if A is the set of real numbers between 0 and 2, and B the set between 1 and 3, $A \cup B$ is the set between 0 and 3.

The *common part*, or intersection, $A \cap B$, of the sets A and B is the set of things belonging to both A and B, i.e. "$x \in A \cap B$" means "$x \in A$ and $x \in B$".

The principal identical relations involving $\cup$ and $\cap$ are

B 1. $A \cup (B \cup C) = (A \cup B) \cup C, \quad A \cap (B \cap C) = (A \cap B) \cap C.$

2. $A \cup B = B \cup A, \quad A \cap B = B \cap A.$

3·1. $A \cap (B \cup C) = (A \cap B) \cup (A \cap C).$

3·2. $A \cup (B \cap C) = (A \cup B) \cap (A \cup C).$

4. $A \cup A = A, \quad A \cap A = A.$

It will be noticed that $\cup$ and $\cap$ behave in many ways like ordinary addition and multiplication, **B** 1, 2 and 3·1 correspond-

* The numbers [1], [2], etc., refer to the notes at the end of the book.

ing to the associative, commutative and distributive laws. In accordance with this analogy, AB will be used as an abbreviation for $A \cap B$, with the bracketing convention $AB \cup C = (AB) \cup C$. Properties B 1–4 may then be summed up in the rules that (a) expressions may be transformed as in ordinary algebra, except that there is no cancelling, i.e. neither of the relations $AB = AC$ and $A \cup B = A \cup C$ implies $B = C$; and (b) "multiples" and "powers" of A are to be replaced by A (B 4). The dual distributive law B 3·2 is deducible from the rest (see Example 1).

The following inclusion relations hold:

B 5. $A \subseteq A \cup B$, $A \supseteq A \cap B$.

  6·1. If $A \subseteq C$ and $B \subseteq C$ then $A \cup B \subseteq C$.

  6·2. If $A \supseteq C$ and $B \supseteq C$ then $A \cap B \supseteq C$.

All the formulae B may be accepted as self-evident as they stand, or referred back to propositions of logic. For example, if the "definitions" of $A \subseteq B$ and of $x \in A \cup B$ are inserted in B 6·1, it becomes "if $x \in A$ implies $x \in C$, and $x \in B$ implies $x \in C$, then $(x \in A$ or $x \in B)$ implies $x \in C$".

If A is a finite set, with members $a, b, ..., k$, then

$$A = (a) \cup (b) \cup ... \cup (k),$$

which will be abbreviated to $\{a, b, ..., k\}$.

Examples. 1. Prove B 3·2 from A and the rest of B 1–6.

$(A \cup B)(A \cup C) = A \cup AB \cup AC \cup BC$, by B 1–4, without 3·2.
Since $AB \subseteq A$, $A \cup AB = A$ by B 6·1, and hence

$$(A \cup AB) \cup AC = A \cup AC = A,$$

giving the result.

2. A necessary and sufficient condition that $A \subseteq B$ is that $A \cup B = B$.

If $A \cup B = B$, $A \subseteq B$ by B 5. If $A \subseteq B$, $A \cup B \subseteq B$ by A 1 and B 6·1, and $B \subseteq A \cup B$ by B 5.

Exercises. 1. Prove B 4 formally from A and the rest of B 1-6 (excluding B 3·2).

2. A necessary and sufficient condition that $A \subseteq B$ is that $AB = A$.

5. The *null-set*, or *empty set*, denoted by 0, has no members and is contained in every set:

C. $0 \subseteq A$: *the null-set is a subset of every set.*

When sets are regarded as collections or heaps of things a set with no members is a rather shadowy or even paradoxical entity, but its mysterious quality disappears if statements about sets are interpreted as statements about properties. Let p be called a *null-property* if it is not possessed by any object. Examples are: being greater than 3 and less than 2, or being a zero of e^x. Such properties are frequently considered in mathematics, particularly in proofs by *reductio ad absurdum*. Any two null-properties are "formally equivalent", in the sense of para. 2, and therefore all these properties determine the same set, which is called the null-set.

To arrive from this definition at proposition C it is necessary to consider more closely the interpretation of the symbol $A \subseteq B$. The meaning assigned to it was: for every x, $x \in A$ implies $x \in B$. This means that $x \in B$ unless "$x \in A$" is false, i.e. "$x \in B$" *is true, or* "$x \in A$" *is false*. This final form may be taken as the basic meaning of $A \subseteq B$, and from it it is clear that $0 \subseteq A$. For since, for all x, "$x \in 0$" *is* false, the proposition

"$x \in A$" is true or "$x \in 0$" is false

is true, whatever the set A may be.

From C and B 6·1 and 6·2 it follows that

$$A \cup 0 = A, \quad A \cap 0 = 0.$$

Thus the formal properties of the null-set justify the symbol 0.

Two sets are said to *meet* (or intersect) if they have at least one common member. It follows from the definition of the null-set that the necessary and sufficient condition that A and B meet is that $AB \neq 0$. Two sets that do not meet are said to be *disjoint*.

Note. In work that is not purely symbolical the symbol $A \subseteq B$ is often replaced by the words "all a's are b's"—for example, "the set of all parabolas is contained in the set of all conics" by "all parabolas are conics". It must, however, be borne in mind that if A is the null-set "all a's are b's" is to be regarded as true whatever B may be. Thus all zeros of e^x are real and positive, because e^z has no zeros. All zeros of e^z are also real and negative, and there is no contradiction between the statements, because it is not asserted that any actual number is both positive and negative, but only that the set of zeros of e^z (i.e. the null-set) is a subset of both the other sets of numbers.

6. If $A \subseteq S$ the set of elements of S not belonging to A is called the *complement*, or *residual set*, of A with respect to S. It is denoted by $S - A$; but if S is supposed fixed, $S - A$ may also be denoted by $\mathscr{C}A$.

   Besides the obvious properties

D 1. $\mathscr{C}S = 0, \quad \mathscr{C}0 = S,$

   2. $A \cup \mathscr{C}A = S, \quad A \cap \mathscr{C}A = 0,$

   3. $\mathscr{C}(\mathscr{C}A) = A,$

   4. *If* $A \subseteq B$ *then* $\mathscr{C}B \subseteq \mathscr{C}A,$

the complement has the important property of interchanging $\cup$ and $\cap$:

D 5. $\mathscr{C}(A \cup B) = \mathscr{C}A \cap \mathscr{C}B, \quad \mathscr{C}(A \cap B) = \mathscr{C}A \cup \mathscr{C}B.$

(See Figs. 1 and 2, where S may be taken to be the whole plane.) This proposition corresponds to the theorem in logic that "not $(p \text{ or } q)$" is equivalent to "not p and not q", and "not $(p \text{ and } q)$" to "not p or not q".

   * Theorem 6·1. *If* $A \cup X = S$, *and* $AX = 0$, *then* $X = \mathscr{C}A$.

   By the first of the given equations, $X \subseteq S$. Multiplying the first equation of **D** 2 by X and using $AX = 0$, we obtain

$$X \cap \mathscr{C}A = X.$$

Multiplying the first of the given equations by $\mathscr{C}A$ and using $A \cap \mathscr{C}A = 0$, we get $X \cap \mathscr{C}A = \mathscr{C}A$. Hence $X = \mathscr{C}A$.

   If A and B are any sets, $B - A$ denotes the set of elements of B not belonging to A (Fig. 3). Thus if complements are formed with respect to a set containing A and B as subsets, $B - A = B\mathscr{C}A$.

---

   * I.e. Theorem 1 of para. 6 (of Ch. I), referred to in this chapter as 6·1, in others as I. 6·1.

Evidently $A - A = 0$ and $A - 0 = A$. However, this "minus" sign does not combine with $\cup$ in the way that the "plus"-analogy might suggest. For example, $(A \cup A) - A = 0$, but $A \cup (A - A) = A$. For this reason it is best to begin formal proofs by replacing differences $B - A$ by the equivalent $B\mathscr{C}A$.

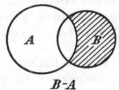

$B\text{-}A$

Fig. 3

Examples. (All complements are formed with respect to an arbitrary set, S, containing A, B and C, except in Example 4.)

1. $A(B - C) = B(A - C) = AB - C$, for all three sets are $AB \cap \mathscr{C}C$.

2. $B - A = B - AB$.
$$
\begin{aligned}
B - AB &= B \cap \mathscr{C}(AB) \\
&= (B \cap \mathscr{C}A) \cup (B \cap \mathscr{C}B) \\
&= B \cap \mathscr{C}A = B - A.
\end{aligned}
$$

3. A necessary and sufficient condition that $AB = 0$ is that $A \subseteq \mathscr{C}B$.

If it is given that $A \subseteq \mathscr{C}B$, multiply both sides by B. If it is given that $AB = 0$, multiply both sides of $A \subseteq B \cup \mathscr{C}B$ by A.

4. If $C \subseteq B \subseteq A$ then $(A - B) \cup (B - C) = A - C$. If complementation is with respect to A, the left-hand side is $\mathscr{C}B \cup B\mathscr{C}C$. Since $\mathscr{C}B \subseteq \mathscr{C}C$, this is
$$
(\mathscr{C}B \cap \mathscr{C}C) \cup B\mathscr{C}C = \mathscr{C}C = A - C.
$$

5. *The propositions* **D**1 *and* **D**3–5 *can be deduced formally from* **A**, **B** *and* **D**2. Since $A\mathscr{C}A \subseteq A$ (by **B**5), **C** follows immediately from **D**2. The theorem 6·1 is proved as in the text. It uses only **A**–**C** and **D**2.

Proof of **D**1. Since $S \cup 0 = S$ and $S0 = 0$, it follows from 6·1 that $\mathscr{C}S = 0$ and $\mathscr{C}0 = S$.

Proof of **D**3. By **D**2 the equations
$$
\mathscr{C}A \cup X = S, \quad \mathscr{C}A \cap X = 0
$$
are satisfied by $X = A$. Therefore by 6·1, $A = \mathscr{C}(\mathscr{C}A)$.

Proof of **D**4. Since $A \subseteq B$, $S = {}'A \cup \mathscr{C}A \subseteq B \cup \mathscr{C}A$. Multiply both sides by $\mathscr{C}B$:
$$
\mathscr{C}B \subseteq \mathscr{C}B \cap \mathscr{C}A \subseteq \mathscr{C}A.
$$

Proof of **D**5. Let $A' = \mathscr{C}A$ and $B' = \mathscr{C}B$. Then
$$
(A \cup B) \cap (A'B') = (AA' \cap B') \cup (BB' \cap A') = 0,
$$
$$
(A \cup B) \cup (A'B') \supseteq (AB' \cup A'B \cup AB) \cup A'B',
$$

since all the sets in the bracket on the right are contained in A or B,

$$= (A \cup A') \cap (B \cup B') = S.$$

Since A, B and $A'B'$ are all contained in S, it follows that

$$(A \cup B) \cup A'B' = S.$$

Hence by 6·1 $A \cup B$ and $A'B'$ are complementary sets. This gives the first part of D 5 immediately, and the second part on interchanging dashed and undashed letters.

Exercises. 1. $A(B-C) = AB - AC$.

2. $\mathscr{C}(A-B) = B \cup \mathscr{C}A$.

3. $(A-B) \cup (A-C) = A - BC$.

4. $(A-C) \cup (B-C) = (A \cup B) - C$.

5. $(A-B) \cup (B-A) = (A \cup B) - AB$.

6. $A - (A-B) = AB$.

7. $\mathscr{C}(A_1 \cup A_2 \cup \ldots \cup A_k) = \mathscr{C}A_1 \mathscr{C}A_2 \ldots \mathscr{C}A_k$.

8. $A \cup (B-A) = A \cup B$, $A(B-A) = 0$. Prove that the whole of D 1–5, and the equation $B - A = B\mathscr{C}A$ can be deduced formally from these two relations together with A, B and the definition "$\mathscr{C}A = S - A$ if $A \subseteq S$". [First prove that the equations

$$A \cup X = A \cup B, \quad AX = 0$$

have at most one solution.]

7. *Duality.* The calculus that has now been developed (Boolean Algebra[2]) has a duality property which has probably already been observed by the reader. If in any theorem of the Algebra all differences are expressed in terms of complements with respect to a fixed set S, and then the symbols

$$\left. \begin{array}{l} \cup \text{ and } \cap \\ 0 \text{ and } S \\ \subseteq \text{ and } \supseteq \end{array} \right\} \quad \text{are everywhere interchanged,}$$

the result is also a true theorem of the Algebra.

Since no appeal is made to the duality property in this book a general proof (which would require a more exact definition of Boolean Algebra) is not given. (Cf. Note 2.)

8. It is frequently necessary to consider sets whose members are themselves sets of things. If **M** is such a set of sets, the members of **M** (i.e. the sets of "things") are usually denoted by a suffix notation, A_x. The suffix x may range through any set B, e.g. the integers from 1 to k, all the positive integers, all the real numbers, etc. When this notation is used for the members of **M** the set **M** itself is denoted by $\{A_x\}$.

The *union*
$$\bigcup_{x \in B} A_x$$

is the set of all members of the sets A_x; i.e.
$$\text{"}z \in \bigcup_{x \in B} A_x\text{"}$$

means "for some x of B, $z \in A_x$". The *intersection*
$$\bigcap_{x \in B} A_x$$

is the set of elements that belong to all the A_x; i.e.
$$\text{"}z \in \bigcap_{x \in B} A_x\text{"}$$

means "for every x of B, $z \in A_x$". The notations for union and intersection may be abbreviated to $\bigcup A_x$ and $\bigcap A_x$, $\bigcup_x A_x$ and $\bigcap_x A_x$, or even $\bigcup A$ and $\bigcap A$, when the meaning is clear. When the suffixes are positive integers the union is denoted by

$$\bigcup_1^k A_n \quad \text{or} \quad \bigcup_1^\infty A_n,$$

and the intersection similarly; but it is to be emphasised that the infinite union and intersection are not derived from the finite ones by any limiting process, but have an independent definition of their own.

The definitions evidently agree with those previously given for "union" and "intersection" when the number of sets is finite.

Example. If A_n is the set of roots of the equation $z^n = 1$, $\bigcup_1^\infty A_n$ is the set of numbers $e^{2\pi i \alpha}$, where α takes all rational values; and $\bigcap_1^\infty A_n$ is the single number 1.

The formal properties of $\cup$ and $\cap$ are:

E 1.   *If* $a \in B$, $\displaystyle\bigcap_{x \in B} A_x \subseteq A_a \subseteq \bigcup_{x \in B} A_x$.

2·1.   *If, for every a of B, $A_a \subseteq C$, then $\cup A_x \subseteq C$.*

2·2.   *If, for every a of B, $A_a \supseteq C$, then $\cap A_x \supseteq C$.*

These propositions may be "translated" in the usual way; e.g. E 2·1 states that if $a \in B$ implies $A_a \subseteq C$, then "$z \in A_a$ and $a \in B$" implies $z \in C$.

F 1.   *If $A_x \subseteq B_x$ for each x, $\cup A_x \subseteq \cup B_x$ and $\cap A_x \subseteq \cap B_x$.*

2. $\cup(A_x \cup B_x) = \cup A_x \cup \cup B_x$.

3·1. $\cap(A \cup B_x) = A \cup \cap B_x$.

3·2. $\cup(AB_x) = A \cup B_x$.

4.   *If S contains all the sets A_x, then $\cup A_x$ and $\cap(S - A_x)$ are complementary sets in S; i.e. if $\mathscr{C}$ denotes the complement in S,* $\mathscr{C}(\cup A_x) = \cap(\mathscr{C}A_x)$.

As a final example of the use of the "calculus" it will now be shewn that the propositions F are formally derivable from A–E. From this and other examples that have been given in this section it follows that all the "standard forms" A–F can be derived formally from A, B (without B 4), D 2 and E.

Proof of F 1. For every a, $A_a \subseteq B_a \subseteq \cup B_x$ and $\cap A_x \subseteq A_a \subseteq B_a$: apply E 2.

Proof of F 2. Since $A_a \subseteq \cup A_x$ and $B_a \subseteq \cup B_x$, $A_a \cup B_a \subseteq \cup A_x \cup \cup B_x$. Therefore by E 2·1

$$\cup(A_x \cup B_x) \subseteq \cup A_x \cup \cup B_x.$$

The other half follows from F 1.

Proof of F 3·1. Let $X = \cap(A \cup B_x)$. Then, by F 1, $A \subseteq X$ and $\cap B_x \subseteq X$, and therefore

$$A \cup \cap B_x \subseteq X.$$

If B_a is one of the sets B_x, $X \subseteq A \cup B_a$, and therefore

$$X - A \subseteq (A \cup B_a) - A \subseteq B_a.$$

Hence $X - A \subseteq \cap B_x$, and therefore $X \subseteq A \cup \cap B_x$.

Proof of **F** 3·2. If $Y = \mathsf{U}(AB_x)$ then, by **F** 1, $Y \subseteq \mathsf{U}B_x$ and $Y \subseteq A$, and therefore $Y \subseteq A \mathsf{U} B_x$. If S is a set containing all the sets involved (e.g. $S = A \cup \mathsf{U}B_x$),

$$B_a = AB_a \cup (S-A) B_a$$
$$\subseteq Y \cup (S-A).$$

Therefore $\mathsf{U}B_x \subseteq Y \cup (S-A),$

and $A \mathsf{U} B_x \subseteq A Y \subseteq Y.$

Proof of **F** 4. Let $\mathsf{U}A_x = X, \bigcap (S-A_x) = Y.$ Then clearly $X \cup Y \subseteq S,$
and
$$X \cup Y = X \cup \bigcap (S-A_x) = \bigcap (X \cup (S-A_x))$$
$$\supseteq \bigcap (A_x \cup (S-A_x)) = S;$$

$$XY = Y \cup A_x = \mathsf{U}(YA_x) \subseteq \mathsf{U}((S-A_x) \cap A_x) = 0.$$

The result now follows by 6·1.

§ 2. ENUMERABLE AND NON-ENUMERABLE SETS

9. A (1, 1)-*transformation*, f, of a set A on to a set B (or a (1, 1)-*correspondence* between the sets) is determined if with every element x of A there is associated an element $f(x)$ of B, called the

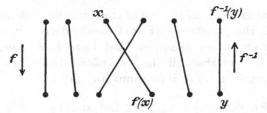

Fig. 4.   (1, 1)-correspondence

image of x, in such a way that each element y of B is the image of just one element of A (which is called $f^{-1}(y)$). The condition is symmetrical between A and B, and f^{-1} is a (1, 1)-transformation of B on to A.

The sets A and B are *similar*, $(A \operatorname{sm} B)$, if a (1, 1)-correspondence can be set up between them. Clearly if $A \operatorname{sm} B$, then $B \operatorname{sm} A$. If $A \operatorname{sm} B$ and $B \operatorname{sm} C$, then $A \operatorname{sm} C$; for if f and g are (1, 1)-transformations of A on to B and B on to C respectively, gf is a (1, 1)-transformation of A on to C.

Examples. A (1, 1)-correspondence is set up between the set of all positive integers, I, and the set of positive even integers, E, by mapping n of I on $2n$ of E; for every member of E is the image of just one member of I. Hence I and E are similar. More generally, if $\phi(n)$ is any strictly increasing single-valued function of the positive integer n, taking only integer values, the set of all the numbers $\phi(n)$ is similar to the set of all positive integers. For from $\phi(n+1) > \phi(n)$ it follows that $\phi(r) \neq \phi(s)$ if $r \neq s$, and therefore ϕ is itself a $(1, 1)$-mapping of the integers on to the set of numbers $\phi(n)$. Thus all such sets as

$$1!,\ 2!,\ 3!,\ \ldots,$$
$$1^1,\ 2^2,\ 3^3,\ \ldots,$$

are similar to $\quad 1,\ 2,\ 3,\ \ldots.$

10. *Finite and infinite.* These words have already been used occasionally as terms needing no explanation, for example on p. 3. The meaning there is that in any particular instance the elements $a, b, \ldots, k$ of A can actually be written down, so that the dots disappear, and it is doubtful if mathematics can dispense with the use, at some stage, of this primitive and undefinable notion of "finiteness". But if the words are to be used in the enunciation of formal theorems some greater precision is necessary.

It is assumed in this book that the ordinary positive integers are well-defined objects, whose fundamental properties (including the Principle of Induction) have been established, or postulated as axioms. *A finite set* is then defined to be one which is either the null-set or, for some positive integer m, similar to the set, I_m, of positive integers not exceeding m. In other words, a finite set, if not null, is one whose members can be named $a_1, a_2, \ldots,$ and a_m. All other sets are *infinite*. Two similar sets are evidently both finite or both infinite.

The following properties of finite sets are assumed implicitly in the simplest everyday use of numbers: every subset of a finite set is finite; the union of a finite set of finite sets is finite; and a finite set cannot be similar to a proper subset of itself. (The third of them is equivalent to the proposition that in whatever order the elements of a finite collection are counted the result is always the same.) Inductive proofs are given in the notes.[3]

11. The set, I, of all positive integers is infinite, for it was shewn in the Example in para. 9 that it is similar to a proper subset of itself (namely, the even positive integers). A set which is similar to I is said to be *enumerably* (or *countably*) *infinite*. A set which is either finite or enumerably infinite is said to be *enumerable* (or *countable*).* A necessary and sufficient condition that a set should be enumerable is therefore that it should be possible to number its members

$$a_1, a_2, a_3, \ldots,$$

where the series may or may not terminate.

It will be shewn in para. 12 that not all sets are enumerable.

Examples. The even positive integers and the members of any strictly increasing sequence of integers have already been shewn to be enumerable.

Theorem 11·1. *Every subset of I is enumerable.*

If the given subset, A, is null the theorem is trivial. If it is not null its members can be re-numbered in order of increasing magnitude, i.e. we take a_1 to be the least member of A (that every non-null subset of I has a least member follows from the Principle of Induction); a_2 to be the least member of $A - (a_1)$, if $A \neq (a_1)$; and in general when $a_1, a_2, \ldots,$ and a_{r-1} have been defined we take a_r to be the least member of $A - (a_1, a_1, \ldots, a_{r-1})$ if that set is not null. Since the number of elements of A less than n cannot exceed $n - 1$, every element of A receives a suffix, i.e. A is in $(1, 1)$-correspondence with I or a set I_n. Hence

Theorem 11·2. *Every subset of an enumerable set is enumerable.*

Theorem 11·3. *The set, I^2, of ordered pairs (m, n), of positive integers is enumerable.*

The "diagonal" method of enumeration is used. First comes $(1, 1)$; then $(2, 1)$ and $(1, 2)$; then the pairs $(3, 1)$, $(2, 2)$ and $(1, 3)$ for which $m + n = 4$, in order of increasing n; and so on. The function correlating I^2 with I is

$$f(m, n) = \tfrac{1}{2}(m + n - 1)(m + n - 2) + n.$$

Corollary. *The positive rational numbers are enumerable,* for the rationals p/q are in $(1, 1)$-correspondence with the ordered pairs (p, q) with no common factor—a subset of I^2.

---

* Some writers confine the use of "enumerable" to sets similar to I.

Theorem 11·4. *The union of an enumerable set*, **M**, *of enumerable sets is enumerable.*

(The members of **M** may overlap, and **M** and its members may be infinite, finite, or null.)

The members of **M** can be denoted by $A_1, A_2, \ldots$, and the members of A_n (if $A_n \neq 0$) by

$$a_{n1}, a_{n2}, \ldots.$$

If x is any member of $\mathsf{U} A_n$ there may be more than one pair of suffixes m, n such that $x = a_{mn}$, but of all these pairs there is one whose correlate, r, in the transformation of I^2 on to I defined in 11·3 is least. We define $g(x)$ to be r. This rule correlates a unique positive integer with each element of $\mathsf{U} A_n$, and different integers with different elements. Therefore $\mathsf{U} A_n$ is similar to a subset of I, and hence, by Theorem 11·1, it is enumerable.

From 11·4 it follows that the set of all integers (positive, negative and zero), and the set, Q, of all rational numbers are enumerable.

Example. Any set of mutually external circles, all lying within a fixed circle in a plane, is enumerable. The total area of any finite subset of the circles is the sum of their areas. Let α be the radius of the circle containing them all, and η any positive number. If m is the greatest integer not exceeding α^2/η^2, not more than m of the circles can have radius greater than η. The set, E_n, of circles of the set with radius greater than $1/n$ is therefore a finite set, and

$$\overset{\infty}{\underset{1}{\mathsf{U}}} E_n,$$

which is the whole set, is enumerable.

Exercises. 1. In the previous example the condition that all the circles lie inside a fixed circle may be omitted.

2. Any set of mutually external closed intervals on a line is enumerable.

3. If $f(\xi)$ is a single-valued real function of the real variable ξ, and the set of all finite sums

$$f(\xi_1) + f(\xi_2) + \ldots + f(\xi_r) \quad \text{(for any } r\text{)}$$

is bounded, the set of points at which $f(\xi) \neq 0$ is enumerable.

4. The points of discontinuity of a monotone function of a real variable form an enumerable set. [Use the previous example, taking $f(\xi)$ to be the leap (oscillation) of the given function at ξ.]

Theorem 11·5. *The set of all ordered finite sets of positive integers is enumerable.*

With the member $(n_1, n_2, ..., n_r)$

of the given set, P, correlate the integer

$$2^{n_1} . 3^{n_2} . .... p_r^{n_r},$$

where p_r is the rth prime. Since factorisation into prime powers is unique this is a $(1, 1)$-transformation of P on to a subset of I. Therefore P is enumerable. Hence

Theorem 11·6. *The set, I^k, of ordered sets of k positive integers is enumerable*, for it is a subset of P.

In 11·5 and 11·6 "ordered sets of positive integers" may evidently be replaced by "ordered finite selections from A (repetitions allowed)", where A is any enumerable set. Taking A to be Q, the set of rational numbers, we find

Theorem 11·7. *The set, Q^k, of rational points (points with rational coordinates) of cartesian k-space, is enumerable.*

Example. (Second proof of the example on p. 13.) In each circle choose a rational point of the plane. These points are all distinct, and therefore in $(1,1)$-correspondence with the circles; and they are a subset of Q^2, an enumerable set.

Exercises. 1. Prove that the set, $I[t]$, of polynomials in t with integral coefficients is enumerable, and deduce that the set of algebraic complex numbers is enumerable.

2. The word "ordered" may be omitted in Theorems 11·5 and 11·6. [Use 11·4.]

12.   Theorem 12·1. *The set R of real numbers is not enumerable.*

Let E be the set of those real numbers, ξ, satisfying $0 \leqslant \xi < 1$, that are expressible as decimals, finite or infinite, by means of 0's and 1's alone. If R is enumerable the subset E is enumerable, say as $\xi_1, \xi_2, ..., \xi_n, ...$

Let α be the decimal whose nth figure is defined, for each n, to be 0 if the nth figure of ξ_n is 1, and 1 if the nth figure of ξ_n is 0. Then α is not equal to ξ_n for any n, for its decimal differs from that of ξ_n in the nth place, and decimal representation by 0's and 1's is unique; but $\alpha \in E$. Thus the initial assumption, that every member of E is a ξ_n, has led to a contradiction.

We have actually shewn that the numbers $0 \leqslant \xi < 1$ have a non-enumerable subset. Therefore, a fortiori

Theorem 12·2. *The set of real numbers* $0 \leqslant \xi < 1$ *is not enumerable.*

13. *Equivalence relations.* A relation, $x R y$, between x and y is said to be an *equivalence relation* if, for all x, y and z,

(a) $x R x$ (R is *reflexive*),

(b) if $x R y$ then $y R x$ (R is *symmetrical*),

(c) if $x R y$ and $y R z$ then $x R z$ (R is *transitive*).

Examples of equivalence relations (between integers) are: (1) having the same remainder on division by 2, and (2) having the same number of distinct prime factors. The relations $x < y$ and $|x - y| < 5$, between integers, and $A \subseteq B$ between sets, are not equivalence relations: the condition (b) fails for $<$ and $\subseteq$, the condition (c) for $|x - y| < 5$.

The fundamental property of an equivalence relation, R, is that it divides the whole set of objects to which it applies into mutually exclusive classes, such that $x R y$ if, and only if, x and y belong to the same class. Another way of stating the same result is that a mark can be attached to each object in such a way that x and y carry the same mark if, and only if, $x R y$. (In Example (1) above there are two classes, and the marks are "odd" and "even".)

The proof of these statements is as follows. For each x let A_x be the set of all elements, y, such that $x R y$. Then $x \in A_x$, by (a). If $x R y$, $A_x = A_y$. For if z is any member of A_y, $y R z$ and, since $x R y$, $x R z$. Thus $z \in A_x$ and hence $A_y \subseteq A_x$. Similarly, since by (b) $y R x$, $A_x \subseteq A_y$. It now follows that if A_s and A_t meet they are identical, for if u is a common member $s R u$ and $t R u$, and therefore $A_s = A_u = A_t$. Thus the classes A_x, ... are mutually exclusive; if x and y both belong to A_z, $A_z = A_x$, and therefore $y \in A_x$, i.e. $x R y$; and conversely if $x R y$, x and y both belong to A_x.

Cardinal numbers. Similarity is an equivalence relation, and therefore a "mark", $\mathfrak{n}(A)$, can be attached to every set in such a way that $\mathfrak{n}(A) = \mathfrak{n}(B)$ if, and only if, A sm B. It is this mark which we call the *cardinal number*, or cardinal, of the set.

Since, if $m < n$, the finite set I_n is not similar to its proper subset I_m, every finite set is similar to one, and only one, of the sets I_m, and the corresponding m is taken to be the cardinal number of the set. The cardinal of I, and therefore of all enumerably infinite sets, is denoted by $\aleph_0$, that of R, the set of real numbers, by c. Theorem 12·1 states that $\aleph_0 \neq c$.

The infinite cardinal numbers, i.e. cardinals of infinite sets, have an arithmetic of their own resembling in many respects that of the finite integers, but it is not needed in this book.[4]

Chapter II

CLOSED SETS AND OPEN SETS
IN METRIC SPACES

§ 1. CLOSED AND OPEN SETS

1. A *metric* is set up in any set by associating with every pair of elements, x and y, a non-negative number $\rho(x, y)$, called the *distance* between them, satisfying the following conditions:

$(m_1)\ \rho(x, y) = \rho(y, x)$,

$(m_2)\ \rho(x, y) = 0$ if, and only if, $x = y$,

$(m_3)\ \rho(x, y) + \rho(y, z) \geqslant \rho(x, z)$.

A non-null set in which a metric has been defined is called a *metric space*, and its elements are called points.

The most familiar metric spaces are the *open cartesian spaces*, R^p. The points of R^p are the ordered sets of p real numbers,† $(\xi_1, \xi_2, ..., \xi_p)$, $(p \geqslant 1)$, and the distance between $x, = (\xi_1, \xi_2, ..., \xi_p)$, and $y, = (\eta_1, \eta_2, ..., \eta_p)$, is by definition

$$|x - y| = \sqrt{\sum_1^p (\xi_i - \eta_i)^2}.$$

That this function satisfies m_1 and m_2 is obvious; and by the Schwarz inequality,

$$|x - y| \cdot |y - z| \geqslant \sum_1^p (\xi_i - \eta_i)(\eta_i - \zeta_i),$$

and therefore

$$(|x - y| + |y - z|)^2 \geqslant \sum_1^p (\xi_i - \eta_i)^2 + \sum_1^p (\eta_i - \zeta_i)^2 + 2\sum_1^p (\xi_i - \eta_i)(\eta_i - \zeta_i)$$

$$= |x - z|^2.$$

Hence m_3 is satisfied. The "origin", $(0, 0, ..., 0)$, is denoted by o.

Another important metric space is the *real Hilbert space*, H^∞, whose points are all the sequences

$$(\xi_1, \xi_2, ...)$$

---

† Real numbers are denoted by Greek letters, to distinguish them from points, denoted by small italic letters.

of real numbers such that $\Sigma \xi_i^2$ is convergent. If x and y are the points of this space with coordinates (ξ_i) and (η_i), and α and β are any real numbers, the sequence $(\alpha \xi_i + \beta \eta_i)$ also defines a point of H^∞, for by the Schwarz inequality

$$\sum_m^n (\alpha \xi_i + \beta \eta_i)^2 \leqslant \alpha^2 \sum_m^n \xi_i^2 + \beta^2 \sum_m^n \eta_i^2 + 2 \,|\, \alpha \beta \,|\, \left(\sqrt{\sum_m^n \xi_i^2} \right) \left(\sqrt{\sum_m^n \eta_i^2} \right).$$

In particular $\Sigma (\xi_i - \eta_i)^2$ is convergent, and we define the distance between x and y to be

$$|\, x - y \,| \,=\, \sqrt{\sum_1^\infty (\xi_i - \eta_i)^2}.$$

The conditions m_1 and m_2 are again obviously satisfied, and

$$|\, x - y \,| + |\, y - z \,| \geqslant \sqrt{\sum_1^p (\xi_i - \eta_i)^2} + \sqrt{\sum_1^p (\eta_i - \zeta_i)^2} \qquad \text{(any } p\text{)}$$

$$\geqslant \sqrt{\sum_1^p (\xi_i - \zeta_i)^2},$$

by the m_3-property of R^p. Since this holds for every p it holds in the limit, i.e. m_3 is satisfied.

If a and b are the points of R^p or of H^∞ with coordinates α_i and β_i, $\lambda a + \mu b$ denotes the point with coordinates $\lambda \alpha_i + \mu \beta_i$. The set of points $a + \tau (b - a)$ is called the *straight line ab* if τ takes all real values, a *ray* issuing from a if τ takes all non-negative values, and the *segment ab* if the range of τ is $0 \leqslant \tau \leqslant 1$.

Any set of points in a space with metric ρ is itself made into a metric space by the function ρ. In particular, any set of points in R^p may be regarded as an independent metric space.

Example. Product space. If $S_1, S_2, \ldots, S_k$ are metric spaces, with metrics $\rho_1, \rho_2, \ldots, \rho_k$, a new space is formed by taking as *points* all ordered sets $(x_1, x_2, \ldots, x_k)$, where $x_i \in S_i$, and as the *distance* $\rho(x, y)$ between $x = (x_1, x_2, \ldots, x_k)$ and $y = (y_1, y_2, \ldots, y_k)$ the sum

$$\rho_1(x_1, y_1) + \rho_2(x_2, y_2) + \ldots + \rho_k(x_k, y_k).$$

The properties (m_1), (m_2), (m_3) for ρ follow easily from the corresponding properties for the ρ_i. The new space is the *product-space*

$$S_1 \times S_2 \times \ldots \times S_k.$$

The relation of $R^1 \times R^1 \times \ldots \times R^1$ (p factors) to the space R^p is considered in Chapter III (para 2, Example 2).

2. A *real vector space* is a set of elements (points) in which any two points, x and y, determine a point $x+y$, their *sum*, and any point x and real number λ determine a point λx, their *product*; provided that the points form an abelian group under the operation of addition, that the multiplication is associative and distributive, and that $1 . x = x$ for every x. (For a fuller statement of these conditions see the notes.[5]) It follows from these assumptions that there is a unique point o, the *origin*, such that $x+o = x$ for every x; that the equation $a+x = b$ for x has a unique solution, which is denoted by $b-a$; and that $\lambda o = 0x = o$, for every real λ and point x. The definitions of straight line, ray, and segment given above for R^p and H^∞ can be taken over without change to any real vector space.

A *complex* vector space differs from a real one only in that the product λx exists for all complex values of λ. In this book "vector space" means "real or complex vector space".

In any vector space a *norm*, $\|x\|$, is a real function of x such that

(n_1) $\|\lambda x\| = |\lambda| . \|x\|$,

(n_2) if $x \neq o$, $\|x\| \neq 0$,

(n_3) $\|x\| + \|y\| \geqslant \|x+y\|$.

It follows, on putting $\lambda = 0$ in (n_1), that $\|o\| = 0$, and on putting $y = -x$ in (n_3), that $\|x\| > 0$ if $x \neq o$. If $\rho(x,y)$ is taken to be $\|x-y\|$, the conditions (n_i) ensure that the conditions (m_j) for a metric are satisfied. A metric space defined in this way is called a *normed vector space*.

R^p and H^∞ are examples of normed vector spaces. It is convenient in these and certain other special cases to denote the norm by $|x|$.

If (n_1) is replaced by the weaker condition

(n_1^0) $\|-x\| = \|x\|$ and $\|o\| = 0$

it still follows that $\|x-y\|$ is a metric. The space R^ω defined in the following example satisfies (n_1^0) but not (n_1).

Example. If all sequences $(\xi_1, \xi_2, \ldots)$ of real numbers are taken as points, and addition of points and multiplication by a real number are defined as in H^∞, a real vector space R^ω is obtained. A metric can be set up by the rule

$$\|x\| = \sum_1^\infty \frac{1}{n^2} |\xi_n|^*,$$

where $x = (\xi_1, \xi_2, \ldots)$, $\rho(x, y) = \|x - y\|$, and for any non-negative α, $\alpha^* = \min(\alpha, 1)$.† That n_1^0 and n_2 are satisfied is obvious, and n_3 follows from the inequality

$$\min(\alpha + \beta, \gamma) \leqslant \min(\alpha, \gamma) + \min(\beta, \gamma)$$

which is satisfied by any three non-negative numbers.‡ The function $\|x\|$ is not a norm in this case, and R^ω is not a normed vector space. (Cf. p. 58.) The distance function in R^ω will be denoted by $|x - y|$.

Exercises. 1. Prove that the following are normed vector spaces.

(a) The points are all the bounded real functions, ϕ, of a real variable ξ. Addition and multiplication by a real number have their ordinary meanings, and§

$$\|\phi\| = l.u.b. |\phi(\xi)|.$$

(b) *Complex Hilbert space, K^∞.* The points are all sequences of complex numbers,

$$z = (\zeta_1, \zeta_2, \ldots)$$

such that $\sum_1^\infty |\zeta_n|^2$ is convergent. Addition of points and multiplication by a complex number are defined as in H^∞, and $\|z\| = \sqrt{\sum_1^\infty |\zeta_n|^2}$.

2. Let V be the vector space of all real continuous functions $\phi(\xi)$, let $\alpha_1, \alpha_2, \ldots$ be an enumeration of the rational numbers and (cf. Example 1) let

$$\|\phi\| = \sum_1^\infty \frac{1}{n^2} |\phi(\alpha_n)|^*.$$

Shew that $\|\phi\|$ satisfies (n_1^0), (n_2), (n_3).

3. If a is a point of a space with the metric ρ, and ϵ any positive number, the set of all points, x, satisfying $\rho(x, a) < \epsilon$ is called a *spherical neighbourhood*, and more particularly an ϵ-neighbourhood of a. It is denoted by $U(a, \epsilon)$, or by $U_\rho(a, \epsilon)$ when it is desirable to put the metric ρ in evidence. Thus the statements $\rho(x, a) < \epsilon$ and $x \in U(a, \epsilon)$ are equivalent.

† $\min(\alpha, \beta, \ldots \kappa)$ is the least of the finite set of real numbers $\{\alpha, \beta, \ldots, \kappa\}$.
‡ If $\gamma \geqslant \alpha$ and $\gamma \geqslant \beta$, then $\min(\alpha, \gamma) + \min(\beta, \gamma) = \alpha + \beta \geqslant \min(\alpha + \beta, \gamma)$. If (say) $\gamma < \alpha$, then $\min(\alpha, \gamma) + \min(\beta, \gamma) \geqslant \min(\alpha, \gamma) = \gamma \geqslant \min(\alpha + \beta, \gamma)$.
§ $l.u.b.$ = least upper bound, $g.l.b.$ = greatest lower bound.

Theorem 3·1. *If b is any point of $U(a, \epsilon)$, there is a positive δ such that*

$$U(b, \delta) \subseteq U(a, \epsilon).$$

Let $\delta = \epsilon - \rho(a, b)$ (a positive number by hypothesis). Then if $x \in U(b, \delta)$,

$$\rho(a, x) \leqslant \rho(a, b) + \rho(b, x)$$
$$< \rho(a, b) + \delta = \epsilon,$$

i.e. $x \in U(a, \epsilon)$.

A set of points E in S is an *open set* if each of its points has a spherical neighbourhood contained in E. Open sets will be denoted by G, G_1, G_a, etc.

Fig. 5

The null set and the whole space are always open sets.

Examples. (In R^1.) The set† $[\xi > 0]$ is open, for if $\alpha > 0$, the neighbourhood $(\frac{1}{2}\alpha, \frac{3}{2}\alpha)$‡ of α is contained in the set. The set $[\xi \geqslant 0]$ is not open, for the origin belongs to the set but has no neighbourhood contained in it.

(In R^2.) $[\xi_1^2 + \xi_2^2 < 1]$ is an open set. For if x is inside the unit circle, at a distance δ from it, all points of $U(x, \delta)$ are inside the circle. The set $[\xi_1^2 + \xi_2^2 \leqslant 1]$ is not open.

It follows immediately from 3·1 that *every spherical neighbourhood $U(a, \epsilon)$ is an open set.*

Theorem 3·2. *The union of any set of open sets is open.*

Let $E = \bigcup_a G_a$. If $x \in E$, $x \in G_a$ for some a. Therefore some $U(x, \epsilon) \subseteq G_a \subseteq E$.

Theorem 3·3. *The common part of a finite number of open sets is open.*

Let G_1, G_2, ..., G_n be the open sets, and x a point of their common part. Then there exist positive numbers ϵ_r such that

$$U(x, \epsilon_r) \subseteq G_r \quad (r = 1, 2, ..., k).$$

If ϵ is the smallest of the ϵ_r, $U(x, \epsilon) \subseteq G_1 G_2 ... G_k$.

† The square brackets enable us to distinguish between the single set $[\xi_1 > 0, \xi_2 > 0]$ in R^2 (a quadrant), and the two sets $[\xi_1 > 0]$, $[\xi_2 > 0]$ (two half-planes); and also permit such notations as $[\xi_1 > 0] \cup [\xi_2 < 0]$.

‡ If $\beta > \alpha$, $<\alpha, \beta>$ denotes the set of numbers $[\alpha \leqslant \xi \leqslant \beta]$, (α, β) the set $[\alpha < \xi < \beta]$, $<\alpha, \beta)$ the set $[\alpha \leqslant \xi < \beta]$, $<\alpha, \infty)$ the set $[\xi \geqslant \alpha]$ and (α, ∞) the set $[\xi > \alpha]$.

Whereas Theorem 3·2 holds for any infinity of open sets, 3·3 cannot be so extended. In fact, the one-point set (a), which is not in general open, is the common part of the open sets

$$U\left(a, \frac{1}{n}\right) \quad (n = 1, 2, \ldots).$$

Theorem 3·4. *A necessary and sufficient condition for a set E to be open is that it be the union of a set of spherical neighbourhoods.*

For, first, if E is open, each point x of E has a neighbourhood $U(x)$ contained in E. Thus $\mathsf{U}_x U(x) \subseteq E$, and clearly $\mathsf{U}_x U(x) \supseteq E$. Secondly, since all neighbourhoods are open, any union of them is open.

A set of open sets, $\{G_x\}$, is called a *base* if every open set is expressible as the union of sets G_x. Theorem 3·4 therefore states that the spherical neighbourhoods form a base. It may similarly be proved that the ϵ-neighbourhoods for all positive rational ϵ form a base.

Theorem 3·5. *The complement of a finite set of points is open.*

Let $a_1, a_2, \ldots, a_k$ be the points of the finite set, and x any point of the complement E. If ϵ is the least of the positive numbers $\rho(x, a_r)$, $U(x, \epsilon) \subseteq E$.

A *real function* $\phi(x)$ is defined in any space by assigning a real number as *value* to each point x of the space. Such a function ϕ is *continuous* at the point a, if, given $\epsilon > 0$, there exists a positive δ such that $|\phi(x) - \phi(a)| < \epsilon$ whenever $x \in U(a, \delta)$, i.e. whenever $\rho(a, x) < \delta$. This is in agreement with the ordinary definition of continuity of a function of p real variables if S is the space R^p.

Exercise. Prove that the coordinates ξ_i of a point x are continuous functions of x in R^p, H^∞ and R^ω.

Theorem 3·6. *A necessary and sufficient condition for ϕ to be continuous is that, for every real α, the sets $[\phi(x) > \alpha]$ and $[\phi(x) < \alpha]$ are open.*

Necessary. If ϕ is continuous and $\phi(a) > \alpha$, there exists a positive δ such that, if $x \in U(a, \delta)$, $|\phi(x) - \phi(a)| < \phi(a) - \alpha$. It follows that $\quad \phi(x) \geqslant \phi(a) - |\phi(x) - \phi(a)| > \alpha.$

Thus all points of $U(a, \delta)$ belong to $[\phi(x) > \alpha]$. Similarly $[\phi(x) < \alpha]$ is open.

Sufficient. Let a be any point and ϵ a positive number. By hypothesis, for some positive δ_1, $U(a, \delta_1) \subseteq [\phi(x) > \phi(a) - \epsilon]$, and for some positive δ_2, $U(a, \delta_2) \subseteq [\phi(x) < \phi(a) + \epsilon]$. Hence if

$$\delta = \min(\delta_1, \delta_2)$$

$$U(a, \delta) \subseteq [\phi(x) > \phi(a) - \epsilon] \cap [\phi(x) < \phi(a) + \epsilon],$$

that is, at every point x of $U(a, \delta)$, $\phi(a) - \epsilon < \phi(x) < \phi(a) + \epsilon$. Hence ϕ is continuous at a.

Theorem 3·7. *If $\phi_1, \phi_2, ..., \phi_k$ are a finite set of continuous real functions in S, the set of points*

$$[\phi_1 > 0, \ \phi_2 > 0, ..., \phi_k > 0]$$

is open.

The case $k = 1$ is contained in 3·6. In the general case

$$[\phi_1 > 0, \ \phi_2 > 0, ..., \phi_k > 0] = [\phi_1 > 0] \cap [\phi_2 > 0] \cap ... \cap [\phi_k > 0],$$

and is therefore an open set by 3·3.

Example. From 3·5 and 3·7 it follows that in R^2 the square region $[|\xi_1| < 1, |\xi_2| < 1]$, the annulus $[0 < \alpha < \xi_1^2 + \xi_2^2 < \beta]$, the "punctured" circular domain $[0 < \xi_1^2 + \xi_2^2 < 1]$ and the half-plane $[\alpha\xi_1 + \beta\xi_2 + \gamma > 0]$ are open; and in R^p the sets $\left[\sum_1^p \xi_i^2 < 1\right]$ and $\left[\sum_1^p \alpha_r \xi_r > \beta\right]$.

Exercises. 1. If G is open $G - ((a_1) \cup (a_2) \cup ... \cup (a_k))$ is open.

2. The "slab" defined by $\alpha_i < \xi_i < \beta_i$ $(i = 1, ..., k)$ in R^ω is open.

4. The distance $\rho(E_1, E_2)$ between two non-null sets of points is defined to be $g.l.b. \ \rho(x, y)$ for $x \in E_1$ and $y \in E_2$. (If E_1 or E_2 is null the distance is not defined.) In particular $\rho(a, E) = g.l.b. \ \rho(a, x)$ for $x \in E$. Clearly $\rho(a, E) = 0$ if $a \in E$, but $\rho(a, E)$ may also be zero if a is not in E. For example, the origin in R^2 is at zero distance from $[\xi_1 > 0]$. The set of points x for which $\rho(x, E) < \epsilon$ is denoted by $U(E, \epsilon)$.

A set of points E is *closed* if it contains all points that are at zero distance from it. Closed sets are denoted by F, F_1, F_a, etc. The whole space and the null set are always closed.

Theorem 4·1. *The set E is closed if, and only if, its complement is open.*

(a) Suppose $\mathscr{C}E$ is open, and let a be such that $\rho(a, E) = 0$. Since any spherical neighbourhood, U, of a contains a point of E,

U is not contained in $\mathscr{C}E$. Therefore a cannot be in the open set $\mathscr{C}E$, and hence is in E. Thus E is closed.

(b) Suppose E is closed. If E is empty, $\mathscr{C}E$ is the whole space, and therefore open. If $E \neq 0$, and if $a \in \mathscr{C}E$, then $\rho(a, E) = \delta > 0$. No point of $U(a, \tfrac{1}{2}\delta)$ belongs to E, i.e. $U(a, \tfrac{1}{2}\delta) \subseteq \mathscr{C}E$. Thus $\mathscr{C}E$ is open.

The following four theorems follow immediately from 4·1 in combination with 3·2, 3·3, 3·5, 3·6, and the formal properties of complementation (ɪ.D 1–5).

Theorem 4·2. *The common part of any set of closed sets is closed.*

Theorem 4·3. *The union of a finite number of closed sets is closed.*

Theorem 4·4. *Every finite set is closed.*

Theorem 4·5. *A necessary and sufficient condition for the real function ϕ to be continuous is that for every real α the sets $[\phi \geqslant \alpha]$ and $[\phi \leqslant \alpha]$ are closed.*

Corresponding to 3·7 there is the stronger theorem:

Theorem 4·6. *If $\phi_1, \phi_2, \ldots$ and $\psi_1, \psi_2, \ldots$ are continuous real functions in S, the set of points*

$$[\phi_1 \geqslant 0,\ \phi_2 \geqslant 0, \ldots,\quad \psi_1 = 0,\ \psi_2 = 0, \ldots]$$

is closed.

(The number of functions may be infinite, and either ϕ's or ψ's may be absent.) If there are no functions ψ and only one ϕ, the theorem is contained in 4·5. In the general case the given set is the intersection

$$\bigcap_i [\phi_i \geqslant 0] \cap \bigcap_j ([\psi_j \geqslant 0] \cap [-\psi_j \geqslant 0])$$

and is therefore closed by 4·2.

It follows from 4·6 that all the ordinary "curves", "surfaces" and other varieties in R^p, determined by a finite number of equations between continuous functions of the coordinates, are closed.

Examples. All straight lines and segments in R^p are closed. The sets $[\xi_1^2 + \xi_2^2 = 1]$, $[\xi_1^2 + \xi_2^2 \leqslant 1]$, the set of points in and on the unit square, the parabola $\xi_2^2 = \xi_1$, are all closed.

The set $[\xi_1^2 + \xi_2^2 = 1, \xi_1 > 0]$ is an example of a set in R^2 which is neither open nor closed.

5. A *linear subset* of a real vector space is a set of points which, if it contains the points x and y, contains the straight line xy. A *convex* set is one which, if it contains the points x and y, contains the segment xy. (Examples of convex sets in R^2 are: a segment, a straight line, the sets of points

$$[\xi_1^2 + \xi_2^2 < 1] \quad \text{and} \quad [\xi_1 > 0, \xi_2 \geqslant 0].)$$

The common part of any set of linear subsets is itself a linear subset, for if x and y belong to the common part the line xy belongs to all the sets, and therefore to their common part. Similarly, the common part of any set of convex sets is convex.

The points $a_1, a_2, ..., a_k$ are *linearly dependent* if real numbers λ_r, not all zero, exist such that

$$\sum_1^k \lambda_r a_r = o.$$

If $b_1, b_2, ..., b_k$ are linearly independent, and a is any point, the set of points

$$a + \sum_1^k \tau_r b_r,$$

where the τ_r take all real values, is called a $[k]$ (or *k-flat*). *It is a linear subset*, for if

$$x = a + \sum_1^k \tau_r b_r \quad \text{and} \quad y = a + \sum_1^k \sigma_r b_r,$$

then $\qquad x + \lambda(y - x) = a + \sum^k (\tau_r + (\sigma_r - \tau_r)\lambda) b_r,$

and is therefore, for every λ, a point of the set.

If in a real vector space there exists a set of p linearly independent points, but no set of $p + 1$, the space has *algebraic dimension p*. If an arbitrarily large number of linearly independent points can be found, the algebraic dimension is infinite.

2

The following results are proved in the notes.[6]

(a) The algebraic dimension of R^p is p.

(b) Any ray or segment or set $[k]$ in R^p is determined by a finite set of linear equations and inequalities in the coordinates.

(c) The only non-null linear subsets of R^p are the sets $[k]$.

From (b) and (c), combined with 4·6 it follows that in R^p rays, segments and all linear subsets (including straight lines) are closed sets. It will be shewn later (Example 4, p. 44), that in any real vector space, rays, segments and sets $[k]$ are closed, but this is not true of all linear subsets (see Exercise 1, p. 34).

The convex sets

$$\left[\sum_1^p \alpha_r \xi_r < \beta\right], \quad \left[\sum_1^p \alpha_r \xi_r > \beta\right]$$

in R^p are open, by 3·7. They are called the two *sides* of the $[p-1]$ $[\Sigma \alpha_r \xi_r = \beta]$. More generally, the set of points

$$\left[\sum_1^p \alpha_{ir} \xi_r < \beta_i, \quad i = 1, 2, ..., k\right]$$

is an open set, and, as the common part of k convex sets, is convex. If not null it is called an *open convex p-cell*.

6. With any set of points E in a space are associated two sets, its *closure* and its *interior*, which are respectively the least closed set containing E, and the largest open set in E.

The closure, $\bar{E}$ or $\mathscr{K}E$, is defined more explicitly to be the common part of all closed sets containing E,

$$\bar{E} = \mathscr{K}E = \bigcap_{E \subseteq F} F.$$

Since there is at least one closed set containing E, namely, the whole space, $\bar{E}$ is well defined, and it follows immediately from the definition that $E \subseteq \bar{E}$, that $\bar{E}$ is closed (by 4·2), and that every closed set containing E contains $\bar{E}$. Thus $\bar{E}$ is indeed the smallest closed set containing E.

Theorem 6·1. *A necessary and sufficient condition for E to be closed is that $\bar{E} = E$.*

For if E is closed it is itself the smallest closed set containing E. From this it follows that, for any E, $\bar{\bar{E}} = \bar{E}$ (or symbolically $\mathscr{K}\mathscr{K} = \mathscr{K}$). Evidently if $E_1 \subseteq E_2$ then $\bar{E}_1 \subseteq \bar{E}_2$.

Theorem 6·2. *A necessary and sufficient condition that a belong to $\bar{E}$ is that $\rho(a, E) = 0$, i.e. that every spherical neighbourhood of a meets E.*

(This property is often taken as the definition of $\bar{E}$.)

Necessary. Let $a \in \bar{E}$, and if possible let $E \cap U(a, \epsilon) = 0$. The closed set $\mathscr{C} U(a, \epsilon)$ contains E, and therefore $\bar{E}$, contrary to the assumption that $a \in \bar{E}$.

Sufficient. Let $\rho(a, E) = 0$ and let F be any closed set containing E. Then a fortiori $\rho(a, F) = 0$ and hence $a \in F$. Thus a is a common point of all closed sets containing E, i.e. $a \in \bar{E}$.

A set of points E is *bounded* if the distances between pairs of points of E have a finite least upper bound, which is then called *the diameter of E*, and denoted by $\Delta(E)$. The null-set counts as bounded, its diameter is 0. *If E is bounded $\bar{E}$ is bounded, and $\Delta(\bar{E}) = \Delta(E)$.* For clearly $\Delta(\bar{E}) \geqslant \Delta(E)$; and if x and y are any points of $\bar{E}$, and x_1 and y_1 are points of E within an assigned distance ϵ of x and y respectively

$$\rho(x, y) \leqslant \rho(x_1, y_1) + \rho(x, x_1) + \rho(y, y_1) \leqslant \Delta(E) + 2\epsilon.$$

Examples. 1. In R^1 the closure of the set of points $A = \{1, \frac{1}{2}, \frac{1}{3}, \ldots\}$ is $A \cup o$. For o is at zero distance from the set, and therefore $A \cup o \subseteq \bar{A}$. If $\alpha \neq o$, and is not in A, there are only a finite number of points of A within $\frac{1}{2} \lfloor \alpha \rfloor$ of α, and its distance from the nearest of them is $\rho(\alpha, A) > 0$. Hence $\bar{A} \subseteq A \cup o$.

In R^2 let $A_1 = [\xi_1^2 + \xi_2^2 < 1]$, $A_2 = [\xi_1^2 + \xi_2^2 \leqslant 1]$, $A_3 = A_2 - (o)$. Then $\bar{A}_1 = \bar{A}_2 = \bar{A}_3 = A_2$. For A_2 is (by 4·5) a closed set containing A_i and therefore $\bar{A}_i$ $(i = 1, 2, 3)$; and every point of A_2 is at zero distance from A_i; therefore $A_2 \subseteq \bar{A}_i$.

If Q^2 is the set of rational points of R^2, $\bar{Q^2} = R^2$.

2. Let A_4 be the set of points $(m/n, 1/n)$ in R^2, for all integral m and all positive integral n (Fig. 6). Then $\bar{A}_4$ *is the union of A_4 and the ξ_1-axis.* For if (α, β) is not on the ξ_1-axis a circle with radius $\frac{1}{2} | \beta |$ contains only a finite number of points of A_4, and therefore a sufficiently small circle shuts them all out, except possibly (α, β) itself. If $(\alpha, 0)$ is a point

of the ξ_1-axis there is a rational number p/q, with denominator greater than ϵ^{-1}, such that $|\alpha - p/q| < \epsilon$; and the distance of $(p/q, 1/q)$ from $(\alpha, 0)$ is positive but less than $\epsilon\sqrt{2}$.

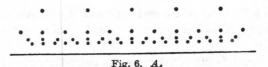

Fig. 6. A_4

Note that if ϕ is a continuous function of the real variables $\xi_1, \xi_2, \ldots, \xi_p$, the set $[\phi \geqslant 0]$ contains, but is not necessarily identical with, the closure of $[\phi > 0]$. For example, the points of R^2 satisfying the inequality

$$\xi_2^2 < \xi_1^2(\xi_1 - 1)$$

all lie to the right of the line $\xi_1 = 1$, but the corresponding equation is satisfied by the origin.

Theorem 6·3. *Given any set of sets* $\{E_a\}$, $\bigcup_a \bar{E}_a \subseteq \overline{\bigcup_a E_a}$, *and* $\overline{\bigcap_a E_a} \subseteq \bigcap_a \bar{E}_a$.

Since, for any admissible b, $E_b \subseteq \bigcup_a E_a$, it follows that

$$\bar{E}_b \subseteq \overline{\bigcup_a E_a},$$

and therefore $\bigcup_b \bar{E}_b \subseteq \overline{\bigcup_a E_a}$. The second half is similarly proved.

Theorem 6·4. $\overline{E_1 \cup E_2 \cup \ldots \cup E_k} = \bar{E}_1 \cup \bar{E}_2 \cup \ldots \cup \bar{E}_k$.

By 4·3, $\bar{E}_1 \cup \bar{E}_2 \cup \ldots \cup \bar{E}_k$ is a closed set, containing

$$E_1 \cup E_2 \cup \ldots \cup E_k.$$

Therefore it contains $\overline{E_1 \cup E_2 \cup \ldots \cup E_k}$.

This, with 6·3, proves the theorem.

This result cannot be extended to the union of an infinity of sets.

Exercises. 1. If A_5 is the set (shewn in Fig. 7)

$$[\xi_2 = \sin(1/\xi_1), \ 0 < \xi_1 \leqslant 1]$$

in R^2, and κ is the segment $[\xi_1 = 0, \ -1 \leqslant \xi_2 \leqslant 1]$, prove that

$$\bar{A}_5 = A_5 \cup \kappa.$$

2. If G is an open set and $EG = 0$, then $\bar{E}G = 0$.

3. $\overline{AB} \subseteq \overline{A} \cap \overline{B}$.

4. If G is open and E any set, $\overline{E}G \subseteq \overline{EG}$. [First prove that if $EG \subseteq F$, then $\overline{E}G \subseteq \overline{F}$.]

5. Prove that the closure of the open convex p-cell (para. 6) is the closed convex p-cell [$\Sigma\alpha_{ir}\xi_r \leqslant \beta_i$, $i = 1, ..., k$]. [Prove that if x is a point of the open cell, and y a point of the closed cell, all points of the segment xy save possibly y belong to the open cell.]

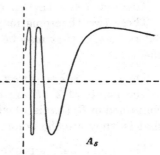

Fig. 7

6. The *convex cover* of any set E in a real vector space is defined to be the common part of all convex sets containing E. It is evidently the "smallest" convex set containing E. Prove that the convex cover of the finite set $(a_1, a_2, ..., a_k)$ is the set of all points

$$\sum_1^k \tau_r a_r,$$

where $\tau_r \geqslant 0$, $\Sigma\tau_r = 1$, (a "$(k-1)$-simplex" if the points a_r are not in any $[k-2]$). [Induction on k.]

7. The *interior* $\mathscr{I}E$ of any set E is the union of all open sets contained in E,
$$\mathscr{I}E = \bigcup_{G \subseteq E} G.$$

Thus $\mathscr{I}E \subseteq E$; by 3·2 $\mathscr{I}E$ is open; and every open set contained in E is contained in $\mathscr{I}E$, which is therefore the greatest open set contained in E. The points of $\mathscr{I}E$ are the *interior points* of E.

Theorem 7·1. *A necessary and sufficient condition for E to be open is that $\mathscr{I}E = E$.* (Compare 6·1.)

It follows that $\mathscr{I}\mathscr{I}E = \mathscr{I}E$ for every E. Evidently if $E_1 \subseteq E_2$, then $\mathscr{I}E_1 \subseteq \mathscr{I}E_2$.

Theorem 7·2. *A necessary and sufficient condition that $a \in \mathscr{I}E$ is that some $U(a) \subseteq E$.*

Necessary. If $a \in \mathscr{I}E$, a belongs to some G contained in E, and for some $U(a)$, $U(a) \subseteq G \subseteq E$.

Sufficient. $U(a)$ is itself an open set. Hence if $U(a) \subseteq E$ then $U(a) \subseteq \mathscr{I}E$, and hence $a \in \mathscr{I}E$.

Theorem 7·3. *Given any set of sets* $\{E_a\}$, $\bigcup_a(\mathscr{I}E_a) \subseteq \mathscr{I}(\bigcup_a E_a)$; *and* $\mathscr{I}(\bigcap_a E_a) \subseteq \bigcap_a(\mathscr{I}E_a)$.

Theorem 7·4. $\mathscr{I}(E_1 \cap E_2 \cap \dots \cap E_k) = \mathscr{I}E_1 \cap \mathscr{I}E_2 \cap \dots \cap \mathscr{I}E_k$.

These two theorems may be proved exactly analogously to 6·3 and 6·4, or they may be deduced formally from the following theorem.

Theorem 7·5. *For any set* E, $\mathscr{C}\mathscr{I}E = \mathscr{K}\mathscr{C}E$ *and* $\mathscr{C}\mathscr{K}E = \mathscr{I}\mathscr{C}E$.

The points of $\mathscr{C}\mathscr{I}E$ are those which have no neighbourhood contained in E; that is, of which every neighbourhood meets $\mathscr{C}E$; that is, they are the points of $\mathscr{K}\mathscr{C}E$.

The second half is proved similarly or follows thus:

$$\mathscr{C}\mathscr{K}E = \mathscr{C}(\mathscr{C}\mathscr{I}\mathscr{C}E) \text{ (by the first part)} = \mathscr{I}\mathscr{C}E.$$

Examples. 1. If A is the set $[\xi_1^2 + \xi_2^2 \leqslant 1]$ in R^2, $\mathscr{I}A$ is $[\xi_1^2 + \xi_2^2 < 1]$. The interior of the circle $[\xi_1^2 + \xi_2^2 = 1]$ is null. (Confusion with ordinary language, according to which the interior of the circle is the set $[\xi_1^2 + \xi_2^2 < 1]$, is best avoided by using the symbol $\mathscr{I}[\xi_1^2 + \xi_2^2 = 1]$ in such cases. In our terminology $[\xi_1^2 + \xi_2^2 < 1]$ is the *inner domain* of the circle (see Chapter v), its points are *inside* the circle.)

2. *If* E *is a convex set in a real vector space with a norm, the sets* $\mathscr{I}E$ *and* $\bar{E}$ *are convex.*

(1) Let a and b be points of $\mathscr{I}E$, and let c be the point

$$\lambda a + (1-\lambda)b \quad (0 < \lambda < 1)$$

of the segment ab. If $\|z\|$ is sufficiently small, the points $a+z$ and $b+z$ lie in spherical neighbourhoods of a and b contained in E; and since

$$c+z = \lambda(a+z) + (1-\lambda)(b+z),$$

$c+z$ also belongs to E in this case. Hence all points in a certain neighbourhood of c belong to E, i.e. $c \in \mathscr{I}E$. Thus $\mathscr{I}E$ is convex.

(2) Let a and b be points of $\bar{E}$, and c the point

$$\lambda a + (1-\lambda)b \quad (0 < \lambda < 1)$$

of the segment ab. By hypothesis there exist points $a+x$ and $b+y$ of E such that $\|x\|$ and $\|y\|$ are less than an assigned positive δ; and the point

$$c' = \lambda(a+x) + (1-\lambda)(b+y) = c + \lambda x + (1-\lambda)y$$

is also in E. By making δ small enough $\|c'-c\|, \leqslant \lambda\|x\| + (1-\lambda)\|y\|$, may be made arbitrarily small. Therefore $c \in \bar{E}$.

Note. From certain of the theorems that have now been proved the reader may be led to suppose that the pairs

closed   and   open

$\mathscr{K}$ and $\mathscr{I}$

could be added to the list of pairs on p. 7, that can be interchanged to give dual theorems. This duality is limited by the asymmetry of the theorem that a single point is a closed set. The definitions of the following section, especially of convergent sequences and compact sets, put an end to the symmetry between "open" and "closed".

8. The *frontier*, $\mathscr{F}E$, of a set E is by definition $\bar{E} - \mathscr{I}E$; we may write symbolically $\mathscr{F} = \mathscr{K} - \mathscr{I}$. Thus

$$\mathscr{F}E = \bar{E} \cap \mathscr{C}\mathscr{I}E = \bar{E} \cap \overline{\mathscr{C}E},$$

by 7·5, shewing that $\mathscr{F}E$ is closed, and that $\mathscr{F}(\mathscr{C}E) = \mathscr{F}E$. It follows that, for any E, the sets $\mathscr{I}E$, $\mathscr{I}(\mathscr{C}E)$ and $\mathscr{F}E$ are three non-intersecting sets whose union is the whole space.

Examples. If A is $[\xi_1^2 + \xi_2^2 < 1]$ in R^2, $\mathscr{F}A$ is the circle $[\xi_1^2 + \xi_2^2 = 1]$. The frontier of the segment $< 0, 1 >$ in R^1 is the pair of points 0 and 1, but the frontier of the segment

$$B: \quad [0 \leqslant \xi_1 \leqslant 1, \ \xi_2 = 0]$$

in R^2 is the segment itself, for $\bar{B} = B$ and $\overline{R^2 - B} = R^2$.

$$\mathscr{F}(R^2 - (o)) = (o).$$

Theorem 8·1. *A necessary and sufficient condition for $\mathscr{F}E = 0$ is that E be both open and closed.*

If E is both open and closed, $\mathscr{F}E = \bar{E} - \mathscr{I}E = E - E = 0$. If $\mathscr{F}E = 0$, then $\bar{E} = \mathscr{I}E$ and hence both are equal to E, which lies between them.

In particular, $\mathscr{F}S$ and $\mathscr{F}(0)$ are always null.

Examples. 1. If $\mathscr{F}E_1$ and $\mathscr{F}E_2$ do not meet, $\mathscr{I}(E_1 \cup E_2) = \mathscr{I}E_1 \cup \mathscr{I}E_2$ and $\mathscr{K}(E_1 \cap E_2) = \mathscr{K}E_1 \cap \mathscr{K}E_2$.

By 7·3 the first equality follows if it is established that

$$\mathscr{I}(E_1 \cup E_2) \subseteq \mathscr{I}E_1 \cup \mathscr{I}E_2.$$

Suppose, if possible, that a belongs to $\mathscr{I}(E_1 \cup E_2)$, but to neither $\mathscr{I}E_1$ nor $\mathscr{I}E_2$. Then there exists a positive ϵ_0, such that, if $\epsilon < \epsilon_0$, $U(a, \epsilon)$ is contained in $E_1 \cup E_2$ but not in E_1 or E_2. This implies that $U(a,\epsilon)$ meets all four of the sets E_1, $\mathscr{C}E_1$, E_2, $\mathscr{C}E_2$; and hence that

$$a \in \bar{E}_1 \overline{\mathscr{C}E_1} \bar{E}_2 \overline{\mathscr{C}E_2} = \mathscr{F}E_1 \mathscr{F}E_2.$$

Since this set is empty there is no such point as a.

The second equality follows on applying the first to $\mathscr{C}E_1$ and $\mathscr{C}E_2$ and using $\mathscr{I}\mathscr{C} = \mathscr{C}\mathscr{K}$.

2. If $\mathscr{F}G_1$ and $\mathscr{F}G_2$ do not meet, $\mathscr{F}(G_1 G_2) = G_1 \mathscr{F}G_2 \cup G_2 \mathscr{F}G_1$. For

$$G_1 \mathscr{F}G_2 \cup G_2 \mathscr{F}G_1 = (G_1 \cup \mathscr{F}G_1)(G_2 \cup \mathscr{F}G_2) - G_1 G_2$$
$$= \bar{G}_1 \bar{G}_2 - G_1 G_2 = \mathscr{F}(G_1 G_2) \quad \text{by Example 1.}$$

Exercises. 1. $\mathscr{F}\bar{E} \subseteq \mathscr{F}E$.

2. $\mathscr{F}(E_1 E_2) \subseteq \bar{E}_1 \mathscr{F}E_2 \cup \bar{E}_2 \mathscr{F}E_1$.

3. For any E_1 and E_2, $\mathscr{I}(E_1 \cup E_2) - (\mathscr{I}E_1 \cup \mathscr{I}E_2) \subseteq \mathscr{F}E_1 \mathscr{F}E_2$, and $\mathscr{K}E_1 \cap \mathscr{K}E_2 - \mathscr{K}(E_1 E_2) \subseteq \mathscr{F}E_1 \mathscr{F}E_2$. [The first was shewn in proving Example 1 above.]

9. The point a is a *limit-point* of the set E if every neighbourhood of a contains an infinity of points of E.[7] The set of limit-points of E is called the *derived set* of E, and denoted by E'. Clearly if $E_1 \subseteq E_2$, $E_1' \subseteq E_2'$. If E is finite, $E' = 0$.

A point may be a limit point of E without belonging to E. For example the set $\{1, \frac{1}{2}, \frac{1}{3}, \ldots\}$ in R^1 has the origin as its only limit-point.

Exercises. Verify that, with the notations of the examples on pp. 27, 28, $A_1' = A_2' = A_3' = A_2$; A_4' is the ξ_1-axis; $A_5' = A_5 \cup \kappa$ and $(Q^2)' = R^2$.

Theorem 9·1. *A necessary and sufficient condition that $a \in E'$ is that $\rho(a, E-a) = 0$, i.e. that every $U(a)$ meets $E-a$.*

Necessary. Since an infinity of points of $U(a)$ are in E, at least one is in $E-a$.

Sufficient. Suppose that a neighbourhood $U(a,\epsilon)$ contains only a finite number of points, $a_1, a_2, \ldots, a_k$ of $E-a$. Then if

$$\delta = \min \rho(a, a_r),$$

$U(a, \delta)$ contains no point of $E-a$, contrary to hypothesis.

Corollary 1. *A necessary and sufficient condition that $a \in E'$ is that every $U(a)$ contains at least two points of E.*

Corollary 2. *For any* E, $\bar{E} = E \cup E'$, *for the statements* "$\rho(a, x) < \epsilon$" and "$x = a$ or $0 < \rho(a, x) < \epsilon$" *are equivalent.*

Corollary 3. *A necessary and sufficient condition for* E *to be closed is that* $E' \subseteq E$, for the condition is equivalent to $E = E \cup E'$. This condition was used as a definition of "closed set" by the earliest writers on point-set theory.

Theorem 9·2. E' *is a closed set.*

Let a be a point of E'' and ϵ a positive number. Within $U(a, \epsilon)$ there is a point b of E', and within $U(b, \epsilon - \rho(a, b))$ an infinity of points of E. Since $U(b, \epsilon - \rho(a, b)) \subseteq U(a, \epsilon)$, $a \in E'$. Thus $E'' \subseteq E'$, and by 9·1, Corollary 3, E' is closed.

Examples already given shew that E'' need not be identical with E', and sets E exist for which the whole series of sets $E', E'', ..., E^{(r)}, ...,$ are different.[8]

The points of $E - E'$ are called the *isolated points* of E, and E is an *isolated set* if all its points are isolated (i.e. if $EE' = 0$). An isolated set is therefore one that contains none of its limit-points. The set $\{1, \frac{1}{2}, \frac{1}{3}, ...\}$ in R^1 and the set A_4 of Example 2, para. 6 are isolated sets.

A set of points E which has no limit point, ($E' = 0$), is called a *discrete set*. Example: the set $\{1, 2, 3, ...\}$ in R^1.

10. A set A is *dense* in a set B containing it if every spherical neighbourhood that contains a point of B also contains a point of A. If B is the whole space S, the condition is simply that every spherical neighbourhood meets A, i.e. that $\bar{A} = S$.

Examples. In the plane R^2, the sets $R^2 - (o)$, the irrational points, the rational points, the points with one rational and one irrational coordinate, are all dense sets.

A set or space is said to be *separable* if it has an enumerable dense subset. The space R^p is separable, since the rational points are enumerable (I. 11·7) and dense.

In a separable space every set of disjoint open sets is enumerable, for each of the open sets (except the null-set, if it is a member) contains a point of the enumerable dense set, and all these points are distinct.

Theorem 10·1. *Every subset of a separable set X is separable.*

Let E be the subset, which we may suppose not null, and $a_1, a_2, \ldots$ an enumerable dense set in X. For each m and n let b_{mn} be a point of E within $\rho(a_n, E) + 1/m$ of a_n. The points b_{mn}, which need not all be distinct, are enumerable. If x is any point of E there is a point a_n within $\frac{1}{3}\epsilon$ of x; $\rho(a_n, E) \leqslant \frac{1}{3}\epsilon$, and therefore, if $m > 3/\epsilon$, $\rho(a_n, b_{mn}) < \frac{1}{3}\epsilon + 1/m < \frac{2}{3}\epsilon$. Hence

$$\rho(b_{mn}, x) \leqslant \rho(a_n, x) + \rho(a_n, b_{mn}) < \epsilon,$$

and the points b_{mn} are dense in E.

Corollary. *Every set in R^p has an enumerable dense subset.*

Exercises. 1. The points of H^∞ with only a finite number of non-zero coordinates are a dense set in H^∞. (Since this is evidently a linear subset it follows that not all linear subsets of H^∞ are closed.)

2. Prove that H^∞ is separable. [The set of points having a finite number of coordinates rational and the rest zero is dense. Use I. 11·5.]

Theorem 10·2. *Every non-enumerable set E in a separable (metric) space S contains a limit point (i.e. $E'E \neq 0$).*

If not, each point x of the set has a neighbourhood $U(x, \delta_x)$ containing no other point of E. No two of the neighbourhoods $U(x, \frac{1}{2}\delta_x)$ can meet, for if $|z - x| < \frac{1}{2}\delta_x$ and $|z - y| < \frac{1}{2}\delta_y$,

$$|x - y| < \tfrac{1}{2}(\delta_x + \delta_y) \leqslant \max(\delta_x, \delta_y),$$

i.e. $x \in U(y, \delta_y)$ or $y \in U(x, \delta_x)$. A dense set in S has a point in each neighbourhood $U(x, \frac{1}{2}\delta_x)$ and is therefore not enumerable.

Theorem 10·3. *If A is a dense set in the metric space S, the neighbourhoods $U(a, \epsilon)$, for all a of A and all positive rational ϵ, form a base.*

Let G be any open set and x a point of G. Then

$$\delta = \rho(x, \mathscr{C}G) > 0,$$

and there is a point a_x of A in $U(x, \frac{1}{3}\delta)$ and a rational number ϵ_x between $\frac{1}{3}\delta$ and $\frac{2}{3}\delta$. Then $x \in U(a_x, \epsilon_x) \subseteq U(x, \delta) \subseteq G$, and therefore $G = \bigcup_x U(a_x, \epsilon_x)$.

Theorem 10·4. *A necessary and sufficient condition for a metric space to be separable is that it have an enumerable base.*

Necessary. This follows immediately from 10·3.

Sufficient. Clearly if the set A is formed by choosing a point in each member of the base, A is both enumerable and dense in S.

Example. In R^p the neighbourhoods $U(a, \epsilon)$, for rational points a and rational ϵ, form an enumerable base; in H^∞ the sets $U(a, \epsilon)$, for rational ϵ and points a having a finite number of coordinates rational and the rest zero.

A set E in S is *nowhere dense* if $S - \bar{E}$ is dense, i.e. if $\mathscr{I}\bar{E} = 0$. A set cannot be both dense and nowhere dense, for if E is dense $S - \bar{E}$ is null. A *closed* set F is nowhere dense if, and only if, it has no interior point.

Examples. 1. The set $\{0, 1, \frac{1}{2}, \frac{1}{3}, \ldots\}$ is nowhere dense in R^1. The segment $[0 \leqslant \xi_1 \leqslant 1, \xi_2 = 0]$ in R^2 is nowhere dense.

2. The graph in R^2 of the (single-valued) continuous function ϕ, i.e. the set

$$\xi_2 = \phi(\xi_1),$$

is nowhere dense. For it is a closed set (4·6) and if it had an interior point—say a point (α_1, α_2) of which an ϵ-neighbourhood belonged to the graph—the whole segment $<\alpha_2 - \frac{1}{2}\epsilon, \alpha_2 + \frac{1}{2}\epsilon>$ of the vertical line $\xi_1 = \alpha_1$ would be contained in the graph, whereas in fact (α_1, α_2) is the only point of the graph on this line.

The condition that ϕ be continuous cannot be omitted (though it may be weakened). For example, let p_n be the nth prime and $(r_1, r_2, \ldots)$ any enumeration of the rational numbers, and if ξ is a rational number of the form m/p_n^q, when expressed in its lowest terms, let $\phi(\xi) = r_n$; for all other values of ξ let $\phi(\xi) = 0$. Then ϕ takes all rational values in every interval, and therefore its graph is dense in the whole plane.

Exercises. 1. A necessary and sufficient condition that E be nowhere dense is that every spherical neighbourhood contain a spherical neighbourhood free of points of E.

2. The graph in R^2 of a monotone function is nowhere dense.

11. A set E is *dense-in-itself* if $E \subseteq E'$, that is if every point of E is a limit-point of the set. A set is *perfect* if $E = E'$, i.e. if E is both dense-in-itself and closed. The famous "Cantor set" described in the example below shews that it is possible for a set of points in R^1 to be both perfect and nowhere dense.

Example. The Cantor set $\mathfrak{C}$ is obtained from the segment $<0,1>$ by removing first the (open) central third $(\frac{1}{3}, \frac{2}{3})$; then the (open) central third of each of the two remaining intervals; then the (open) central third of each of the four remaining intervals; and so on. The removed set, being the union of open intervals, is open, and hence $\mathfrak{C}$ *is closed.* The set $\mathfrak{C}$ may also be specified as the set of all numbers in $<0,1>$ expressible as ternary fractions ("decimals" in the scale of 3) composed of 0's and 2's, i.e. all numbers

$$\sum_1^\infty \frac{a_n}{3^n}, \quad a_n = 0 \text{ or } 2.$$

For the first of the "black intervals" removed in the original definition consists of all ternary fractions $\cdot 1\, c_2 c_3 \ldots$ (except $\cdot 1000 \ldots = \cdot 0222 \ldots$, and $\cdot 1222 \ldots = \cdot 200 \ldots$); the next two black intervals $(\frac{1}{9}, \frac{2}{9})$ and $(\frac{7}{9}, \frac{8}{9})$ contain the fractions $\cdot 01\, c_3 c_4 \ldots$ and $\cdot 21\, c_3 c_4 \ldots$, and so on. At stage n the numbers are removed that have (necessarily) 1 in place n, but not earlier.

Fig. 8

(1) $\mathfrak{C}$ *is dense in itself.* Let $\xi \in \mathfrak{C}$, and let ξ_n be formed from ξ by changing the nth figure of its ternary fraction from 0 to 2 or from 2 to 0. Then all the ξ_n are unequal, and $\xi_n \in U(\xi, 3^{-m})$ if $n > m$.

(2) $\mathfrak{C}$ *is nowhere dense.* If $0 \leqslant \eta \leqslant 1$, let η_n have the same ternary fraction as η save that its nth and $(n+1)$th digits are 1. Then η_n is not in $\mathfrak{C}$, but $\eta_n \in U(\eta, 3^{-n+1})$.

The "measure" of $<0,1> - \mathfrak{C}$ (i.e. the sum of the lengths of its constituent intervals) is 1; but dense open sets in $<0,1>$ exist with arbitrarily small measure (Exercise 2).

Exercises. 1. $\mathfrak{C}$ is a non-enumerable set of points. [Cf. the proof of I. 12·1.]

2. Suppose $0 < \epsilon_n < 1$ for each n, and let X be constructed similarly to $\mathfrak{C}$, except that the "black intervals" removed at stage n bear the ratio ϵ_n to the "white" intervals from which they are removed. Prove that X is closed, and that (1) and (2) hold for X, and deduce that a dense open set in $<0,1>$ can have arbitrarily small measure.

§2. CONVERGENT SEQUENCES OF POINTS. COMPACT SETS

12. Beside the fundamental definitions of closed and open sets there must now be set that of a convergent sequence of points, a direct generalisation of the ordinary convergent sequence of real numbers.

A *sequence* $x_1, x_2, \ldots, x_n, \ldots$, or (x_n), of points of a space is determined by assigning a point x_n to every positive integral value of n. This concept is to be distinguished from that of the mere set of points occurring in the sequence, which is denoted by $\{x_n\}$.

Example. 1. (In R^1.) (i) $x_1 = 0$, $x_n = 1$ for $n > 1$: the sequence $0, 1, 1, 1, \ldots$;

(ii) $y_1 = 1$, $y_n = 0$ for $n > 1$: the sequence $1, 0, 0, 0, \ldots$;

(iii) $z_n = 0$ if n is even, $= 1$ if n is odd: the sequence $1, 0, 1, 0, \ldots$. Here $\{x_n\} = \{y_n\} = \{z_n\}$ but the sequences are distinct.

A sequence of points (x_n) of S *converges* to the point a, in symbols
$$x_n \to a,$$
if, given any positive ϵ, $\rho(x_n, a) < \epsilon$ for all but a finite number of values of n. The notation $\underset{n}{\to}$ is used when other suffixes are present, e.g.
$$x_{mn} \underset{n}{\to} a_m.$$
The condition that $x_n \to a$ can also be written
$$\lim_{n \to \infty} \rho(x_n, a) = 0,$$
where the limit has its ordinary arithmetical sense. Clearly if $x_n = a$ for all but a finite number of n, $x_n \to a$.

Example. 2. Of the three sequences in Example 1, $x_n \to 1$, $y_n \to 0$, and (z_n) does not converge. Since the sets $\{x_n\}$, $\{y_n\}$ and $\{z_n\}$ are the same, this shews clearly that "point of convergence" and "limit-point" are different concepts.

If $x_n \to a$, the set of points $\{x_n\}$ is bounded; for if $x_n \in U(a, \epsilon)$ when $n > n_0$, and if $\eta = \max \rho(x_n, a)$ for $n \leqslant n_0$, then $\rho(x_n, a) < \eta + \epsilon$ for every n.

If $x_n \to a$ and $y_n \to b$, then

$$\rho(x_n, y_n) \to \rho(a, b).$$

For

$$|\rho(x_n, y_n) - \rho(a, b)| \leqslant |\rho(x_n, y_n) - \rho(x_n, b)| + |\rho(x_n, b) - \rho(a, b)|$$
$$\leqslant \rho(y_n, b) + \rho(x_n, a) \to 0.$$

It follows that *a sequence cannot converge to two different points.* For if $x_n \to a$ and $x_n \to b$,

$$\rho(a, b) = \lim \rho(x_n, x_n) = 0.$$

In the space R^1 the condition that the points $\xi_1, \xi_2, \ldots$ converge to the point α is identical with the condition that the numbers $\xi_1, \xi_2, \ldots$ converge, in the ordinary arithmetical sense, to α. It is therefore unnecessary to distinguish between the two, and we write $\xi_n \to \alpha$ for either.

In R^p a necessary and sufficient condition that the sequence of points x_n, with coordinates ξ_{nr}, $(r = 1, \ldots, p)$, should converge to the point a, with coordinates α_r, is that for each r

$$\lim_{n \to \infty} \xi_{nr} = \alpha_r.$$

For if $|x_n - a| \to 0$, a fortiori $|\xi_{nr} - \alpha_r| \underset{n}{\to} 0$; and if for $r = 1, 2, \ldots, p, \xi_{nr} \underset{n}{\to} \alpha_r$, then $|x_n - a| \to 0$.

In H^∞ it is necessary, but not sufficient, that for each fixed r, $\xi_{nr} \to \alpha_r$. Example: the sequence of points $e_1, e_2, \ldots$, where e_n has its nth coordinate 1 and all the rest 0. For each fixed r, $\xi_{nr} \underset{n}{\to} 0$, but $|e_n| = 1$ for every n, and therefore $e_n \nrightarrow o$.

Example. 3. *In R^ω* (p. 19), *if x_n is $(\xi_{n1}, \xi_{n2}, \ldots)$, the condition $\xi_{nr} \underset{n}{\to} \alpha_r$ is necessary and sufficient for $x_n \to a$.*

Necessary: for each fixed r, if n is such that $|x_n - a| < \epsilon < 1/r^2$, then $|\xi_{nr} - \alpha_r|^* < \epsilon r^2 < 1$, and hence $|\xi_{nr} - \alpha_r| < \epsilon r^2$.

Sufficient: given $0 < \epsilon < 1$, suppose r_0 such that

$$\sum_{r_0}^{\infty} \frac{1}{r^2} < \tfrac{1}{2}\epsilon,$$

and n_0 such that $|\xi_{nr} - \alpha_r| < \epsilon/2r_0$ if $n > n_0$ and $r < r_0$. Then if $n > n_0$,

$$|x_n - a| \leqslant \sum_{1}^{r_0 - 1} \frac{1}{r^2} \frac{\epsilon}{2r_0} + \sum_{r_0}^{\infty} \frac{1}{r^2} < \epsilon.$$

Exercises. 1. In the *Hilbert cube*, i.e. the set

$$I^\omega: \quad \left[|\xi_r| \leqslant \frac{1}{r}, \quad r = 1, 2, \ldots \right]$$

in H^∞, the condition $\xi_{nr} \overrightarrow{n} \alpha_r$ is sufficient for $x_n \to a$.

2. A necessary and sufficient condition for $x_n \to a$ in any space is that if G is an open set containing a, $x_n \in G$ for all but a finite number of n.

Theorem 12·1. *A necessary and sufficient condition that $x \in \bar{E}$ is that x be the limit of a convergent sequence, (x_n), of points of E.*

Necessary. If $x \in \bar{E}$ there is a point x_n of E in $U(x, 1/n)$ for every n; and clearly $x_n \to x$.

Sufficient. Since $\rho(x, x_n) \to 0$, $\rho(x, E) = 0$ and $x \in \bar{E}$.

Theorem 12·2. *A necessary and sufficient condition that $x \in E'$ is that x be the limit of a convergent sequence, (x_n), of distinct points of E.*

Necessary. Let x_1 be any point of E other than x, and let $\epsilon_1 = \rho(x, x_1) > 0$. Supposing ϵ_r and x_r defined, let x_{r+1} be a point of $E - (x)$ in $U(x, \epsilon_r)$, and $\epsilon_{r+1} = \min(1/(r+1), \rho(x, x_{r+1}))$. It is easily seen, inductively, that the points $x_1, x_2, \ldots, x_r$ are outside $U(x, \epsilon_r)$ and hence x_{r+1} is different from all of them; and clearly $x_n \to x$.

Sufficient. For any positive ϵ, and a suitable n_0, the points $x_n (n > n_0)$, which are all distinct, lie in $U(x, \epsilon)$. Hence $x \in E'$.

13. A sequence (x_n) is a *fundamental sequence* if

$$\lim_{m, \, n \to \infty} \rho(x_m, x_n) = 0,$$

i.e. if, given $\epsilon > 0$, there exists n_0 such that $\rho(x_m, x_n) < \epsilon$ if $m \geqslant n_0$ and $n \geqslant n_0$. *The points of a fundamental sequence form a bounded set;* for if n_0 is such that $\rho(x_n, x_{n_0}) < 1$ if $n \geqslant n_0$, and if

$$K = \max \rho(x_n, x_{n_0}) \quad \text{for} \quad n < n_0,$$

then $x_n \subseteq U(x_{n_0}, 1 + K)$ for every n.

Theorem 13·1. *Every convergent sequence (x_n) is a fundamental sequence.*

Suppose $x_n \to a$, and let n_0 be such that $\rho(x_n, a) < \frac{1}{2}\epsilon$ if $n > n_0$. Then if also $m > n_0$ we have

$$\rho(x_m, x_n) \leqslant \rho(x_m, a) + \rho(x_n, a) < \epsilon.$$

The converse of 13·1 is not true in all metric spaces. For example, if the open interval $(0, 1)$ of R^1 is regarded as a metric space, the fundamental sequence $(1, \frac{1}{2}, \frac{1}{3}, ...)$ does not converge to any point *of the space*; and in Q^1 (the set of rationals) the fundamental sequence (ξ_n), where $\xi_n = \sum_1^n 1/r!$, is not convergent.

The sequence (y_n) is a *subsequence* of (x_n) if $y_n = x_{r_n}$, where $r_1 < r_2 < r_3 < ....$ Thus in Example 1 of para. 12 (x_n) and (y_n) are subsequences of (z_n), but (x_n) is not a subsequence of (y_n). Clearly if $x_n \to a$, every subsequence of (x_n) converges to a.

Theorem 13·2. *If a fundamental sequence (x_n) has a subsequence converging to a, the whole sequence converges to a.*

Given a positive ϵ, let n_0 be such that $\rho(x_m, x_n) < \frac{1}{2}\epsilon$ if $m, n \geqslant n_0$. Let n_1 be an integer exceeding n_0 such that x_{n_1} is in the subsequence, and $\rho(x_{n_1}, a) < \frac{1}{2}\epsilon$. Then if $n \geqslant n_1$,

$$\rho(x_n, a) \leqslant \rho(x_n, x_{n_1}) + \rho(x_{n_1}, a) < \epsilon.$$

14. The set E is *compact* if every sequence of points in E has a subsequence that converges to a point of E.[9]

Examples. 1. The set of points $\{0, 1, \frac{1}{2}, \frac{1}{3}, ...\}$ in R^1 is compact. For every sequence formed of its points either contains a point, ξ, an infinity of times, and then $(\xi, \xi, ...)$ is a convergent subsequence; or it contains a subsequence of $(0, 1, \frac{1}{2}, ...)$, which converges to 0.

2. The spaces R^p and H^∞ are not compact; for no subsequence of the sequence of points $x_n = (n, 0, 0, ...)$ is bounded, and therefore no subsequence is convergent.

If the set of points E is compact it is closed. If x is any point of $\bar{E}$, there is a sequence, $x_1, x_2, ...,$ of points of E converging to x, and a subsequence converges to a point of E. But every subsequence of $x_1, x_2, ...$ converges to x, which is therefore a point of E.

A closed set in a general space need not be compact (R^p is itself a closed set in R^p); but *every closed set, F, in a compact space*

S *is compact,* for every sequence of points in F has a subsequence converging to a point of S, and therefore of F. Thus in compact spaces closed sets and compact sets are identical.

From 13·2 it follows that *a fundamental sequence in a compact set converges to a point of that set.*

A compact space is bounded. If not, let x_1 be any point of the space, and let x_n be defined inductively as a point such that

$$\rho(x_1, x_n) > \rho(x_1, x_{n-1}) + 1.$$

Then if $r \neq s$, $\rho(x_r, x_s) \geqslant |\rho(x_1, x_r) - \rho(x_1, x_s)| \geqslant 1$:

the sequence $x_1, x_2, \ldots$ has no convergent subsequence, and therefore the space is not compact.

Theorem 14·1. *If the non-null sets F_1 and F_2 in S are compact there exist points a_1 of F_1 and a_2 of F_2 such that $\rho(a_1, a_2) = \rho(F_1, F_2)$.*

For there exists a sequence of pairs of points x_n, y_n, of F_1 and F_2 respectively, such that $\lim \rho(x_n, y_n) = \rho(F_1, F_2)$. There is a convergent subsequence $x_{m_1}, x_{m_2}, \ldots$ of the x_n's such that $x_{m_r} \to a_1$, a point of F_1; and a subsequence $y_{n_1}, y_{n_2}, \ldots$ of the y_{m_r} such that $y_{n_r} \to a_2$ of F_2; and

$$\rho(F_1, F_2) = \lim \rho(x_{n_r}, y_{n_r}) = \rho(a_1, a_2).$$

An immediate corollary is that *if F_1 and F_2 do not meet*

$$\rho(F_1, F_2) > 0.$$

This is still true if only one of the sets, say F_1, is assumed to be compact provided the other is closed. For if x_n, y_n, x_{m_r} and a_1 are as above,

$$\rho(F_1, F_2) = \lim \rho(x_{m_r}, y_{m_r})$$
$$= \lim \rho(a_1, y_{m_r})$$
$$\geqslant \rho(a_1, F_2) > 0,$$

since, by hypothesis, a_1 is not in F_2. If neither set is compact $\rho(F_1, F_2)$ may be zero though both sets are closed. Example: F_1 is the ξ_1-axis, F_2 the set $[\xi_2 = e^{-\xi_1}]$ in R^2.

Theorem 14·2. *A necessary and sufficient condition for E to be compact is that every infinite subset of E have at least one limit-point in E.*

Necessary. Let any infinite subset, X, of E contain the distinct points $x_1, x_2, \ldots$; and let the subsequence (x_{n_r}) of (x_n) converge to a of E. By 12·2, a is a limit-point of $\{x_{n_r}\}$, and a fortiori of X.

Sufficient. Suppose E satisfies the "infinite-subset" condition, and let $x_1, x_2, \ldots$, be any sequence of points of E. If an infinity of the (x_n) are identical, say $x_{n_1} = x_{n_2} = \ldots$, this is the required convergent subsequence. If not, $\{x_n\}$ is an infinite subset of E: let a be a limit-point in E. Let x_{n_1} be any point of $\{x_n\}$ in $U(a, 1)$, and assume inductively that $x_{n_2}, \ldots, x_{n_{r-1}}$ have been defined. Let n_r be the least integer, exceeding n_{r-1}, such that $x_{n_r} \in U(a, 1/r)$. Such an integer exists since $\{x_n\} \cap U(a, 1/r)$ is infinite. This completes the inductive definition, and the subsequence (x_{n_r}) of (x_n) converges to a of E.

Theorem 14·3. *A necessary and sufficient condition for a set of points in R^p to be compact is that it be bounded and closed.*

The necessity of the condition has been proved in the preceding paragraphs. The sufficiency is (by 14·2) equivalent to the famous theorem of Bolzano and Weierstrass, the earliest in the history of the theory of sets of points, that every bounded infinite set of points in R^p has a limit-point.

In proving Theorem 14·3 we shall use, besides the formal arithmetic properties of real numbers, only the theorem that a bounded monotone sequence of real numbers converges to a limit.

(1) $p = 1$. Let the given set, F, be contained in the interval $<\alpha_1, \beta_1>$, and let (ξ_n) be any sequence in F. Let $\xi_{n_1} = \xi_1$. If $\gamma = \frac{1}{2}(\alpha_1 + \beta_1)$, at least one of the half-intervals $<\alpha_1, \gamma>$, $<\gamma, \beta_1>$ contains ξ_n for an infinity of n; choose one such half and call it $<\alpha_2, \beta_2>$, and let $\xi_{n_2} \in <\alpha_2, \beta_2>$, where $n_2 > n_1$. At least one half, $<\alpha_3, \beta_3>$, of $<\alpha_2, \beta_2>$ contains ξ_n for an infinity of n, and therefore a point ξ_{n_3}, where $n_3 > n_2$; and so on. Since, for each n

$$<\alpha_n, \beta_n> \subseteq <\alpha_{n-1}, \beta_{n-1}>,$$

$\alpha_n \geqslant \alpha_{n-1}$ and $\beta_n \leqslant \beta_{n-1}$; and for all n, $\alpha_n < \beta_1$ and $\beta_n > \alpha_1$. Hence (α_n) and (β_n) are bounded monotone sequences of real numbers, and are therefore convergent. Since $|\alpha_n - \beta_n| = 2^{-n+1}|\alpha_1 - \beta_1|$, they converge to the same limit, λ. For any positive ϵ the interval

$(\lambda - \epsilon, \lambda + \epsilon)$ contains $<\alpha_r, \beta_r>$, and therefore ξ_{n_r}, for all sufficiently large r. Therefore the subsequence $\xi_{n_r} \to \lambda$, and since F is closed, $\lambda \in F$.

(2) $p > 1$. If F is bounded, then for each fixed r the ξ_r-coordinates of points of F are bounded, since $|\xi_r| \leqslant |x|$. Therefore if $x_1, x_2, \ldots$ is any sequence of points of F, there is, by the case $p = 1$, a subsequence $x_1^1, x_2^1, \ldots$ whose ξ_1-coordinates converge, say to α_1; there is a subsequence $x_1^2, x_2^2, \ldots$ of this subsequence whose ξ_2-coordinates also converge, say to α_2; and so, finally, there is a subsequence $x_1^p, x_2^p, \ldots$ such that for each r in the range $1, 2, \ldots, p$ the ξ_r-coordinates of the x_n^p converge to a limit α_r. Hence if $a = (\alpha_1, \alpha_2, \ldots, \alpha_p)$, $x_n^p \underset{n}{\to} a$; and since F is closed, $a \in F$.

From the Weierstrass Theorem it follows that R^p is *locally compact*, i.e. every point has a neighbourhood whose closure is compact. For the closure of any neighbourhood $U(x, \epsilon)$ is a closed bounded set, and therefore compact. H^∞ is not locally compact, for (with the notation of p. 38) the sequence

$$x + \epsilon e_1, \ x + \epsilon e_2, \ \ldots$$

lies in the neighbourhood $U(x, 2\epsilon)$ of x, but has no convergent subsequence. This example also shews that the Weierstrass Theorem does not hold in H^∞.

Theorem 14·4. *Every fundamental sequence in R^p is convergent.*

The sequence is bounded, i.e. is contained in the closed set $[|x - a| \leqslant K]$ for some a and K. Hence, by the Weierstrass Theorem and 13·2, it is convergent.

The case $p = 1$ of this theorem, together with 13·1, is the "General Principle of Convergence": a necessary and sufficient condition for the sequence of real terms (ξ_n) to be convergent is that, given $\epsilon > 0$, $|\xi_m - \xi_n| < \epsilon$ when m and n exceed a certain $n_0(\epsilon)$.

It will be shewn in § 4 that 14·4 holds also in H^∞, although the Weierstrass Theorem does not hold there.

Examples. 3. The unit $(p-1)$-sphere in R^p is compact.

4. *The Hilbert cube I^ω* (Exercise 1, p. 39) *is compact.* As the common part of the closed sets $[|\xi_1| \leqslant 1], [|\xi_2| \leqslant \frac{1}{2}], \ldots, I^\omega$ is closed in H^∞. Let $x_1, x_2, \ldots$ be any sequence of points of I^ω. If x_n is $(\xi_{n1}, \xi_{n2}, \ldots)$, then

for each fixed r the coordinates (ξ_{nr}) are a bounded set on the ξ_r-axis. Therefore we may construct (as in the proof of the Weierstrass Theorem) first a subsequence $x_1^1,\, x_2^1,\, ...$, whose ξ_1-coordinates converge, say to α_1; then a subsequence $x_1^2,\, x_2^2,\, ...$, of this subsequence, whose ξ_2-coordinates also converge to α_2; and so on. It is easily seen that the sequence

$$x_1^1,\, x_2^2,\, x_3^3,\, ...$$

is a subsequence of (x_n), and has the property that for each fixed r the ξ_r-coordinates converge to α_r; and therefore $x_n^n \to a$ (Exercise 1, p. 39).

5. *Every compact subset X of H^∞ is nowhere dense.* For if X contains $U(x, \epsilon)$,

$$\overline{U(x, \epsilon)} \subseteq \overline{X} = X,$$

and therefore $\overline{U(x, \epsilon)}$, as a closed subset of a compact set, is compact. This contradicts the result of p. 43, that H^∞ is not locally compact.

6. *The sets $[k]$ ("k-flats") in any normed real vector space are closed.* Consider the set E^k, consisting of all points

$$a + \sum_1^k \tau_r b_r,$$

where the b_r are a set of linearly independent points, and the τ_r take all real values.

The theorem is trivial if $k = 0$. Suppose then that it is true for $k - 1$. From this assumption we deduce the following

Lemma: *There exists a positive number λ (depending on the b_r but not on the τ_r) such that if*

$$x = \sum_1^k \tau_r b_r$$

then $|\tau_r| < \lambda \|x\| \quad (r = 1, 2, ..., k).$

Proof. Since the points b_r are linearly independent, the point b_1 is not expressible in the form

$$\sum_2^k \tau_r b_r.$$

Therefore, if E_0^{k-1} is the set of all such points (a linear subset, therefore closed, by the inductive hypothesis), b_1 is at a positive distance β_1 from E_0^{k-1}. Hence, for all real τ_r,

$$\left\| -b_1 + \sum_2^k \tau_r b_r \right\| \geqslant \beta_1.$$

Multiplying both sides by $|\tau_1|$ it follows, since $\tau_2, \tau_3, \ldots, \tau_k$ are already arbitrary, that

$$\left\| \sum_1^k \tau_r b_r \right\| \geqslant \beta_1 |\tau_1|,$$

for all values of $\tau_1, \tau_2, \ldots, \tau_k$. Similar inequalities hold for $\tau_2, \tau_3, \ldots, \tau_k$, and therefore the lemma follows, on taking λ to be the greatest of the numbers $1/\beta_r$.

Now let x be any point of $\overline{E^k}$, and $x_1, x_2, \ldots$ a sequence of points of E^k converging to x. Given any positive ϵ,

$$\| x - x_n \| < \tfrac{1}{2}\epsilon \quad \text{and} \quad \| x - x_m \| < \tfrac{1}{2}\epsilon,$$

for almost all m and n, and therefore

$$\| x_m - x_n \| < \epsilon,$$

if $m, n > n_0$. Hence, by the lemma, if

$$x_n = a + \sum_{r=1}^k \tau_{nr} b_r,$$

then for any fixed r, $|\tau_{mr} - \tau_{nr}| < \lambda\epsilon,$

if $m, n > n_0$. Since ϵ is arbitrary it follows from 14·4 (case $p = 1$) that, for each r, τ_{nr} converges to a number σ_r. If

$$y = a + \sum_1^k \sigma_r b_r,$$

then $\| x_n - y \| \leqslant \sum_1^k |\tau_{nr} - \sigma_r| \, \| b_r \| \xrightarrow[n]{} 0.$

Hence $x_n \to y$, and therefore $y = x$, i.e. x is a point of E^k.

Exercises. 1. Prove that the convex cover (Exercise 6, p. 29) of a finite set of points in a normed real vector space is compact.

2. The set $[\,|\xi_r| \leqslant 1, \; r = 1, 2, \ldots]$ in R^ω is compact.

15. An ϵ-*net* in E is a finite set of points, A, such that every point of E is within a distance ϵ of at least one point of A; i.e. such that

$$E \subseteq \bigcup_{x \in A} U(x, \epsilon).$$

Theorem 15·1. *A compact set E has an ϵ-net for every positive ϵ.*

Let x_1 be any point of the set, and let x_n be defined inductively to be a point of E distant at least ϵ from each of $x_1, x_2, \ldots, x_{n-1}$, if such a point exists. If there were a point x_n for every n, $\{x_n\}$

would be a sequence of points with no convergent subsequence. Therefore for some n, x_{n-1} exists but not x_n; i.e. all points of the space are within ϵ of $x_1, x_2, \ldots$ or x_{n-1}.

Corollary. *All compact sets are separable*, for if A_n is a $1/n$-net, $\overset{\infty}{\underset{1}{\bigcup}} A_n$ is enumerable and dense in S.

A *covering* of a set E in S is a set of sets $\{A\}$ in S, all meeting E, such that
$$E \subseteq \cup A.$$

If the sets are open, closed, or finite in number, the covering is called open, closed, or finite respectively. It is an *ϵ-covering* if each of the sets has diameter not exceeding ϵ.

The theorem that has just been proved (15·1) states that given ϵ, a finite covering of a compact set can be selected from the set of all ϵ-neighbourhoods. A fundamental theorem, usually called the Heine-Borel Theorem, but given its general form by Lebesgue, states that a finite covering of a compact set F can be selected from *any* open covering of F. A preliminary to the proof of this theorem is the following lemma:

15·2. (Lemma.) *If $\{G_z\}$ is an open covering of the compact space S, there exists a positive number ϵ such that every ϵ-neighbourhood is contained in one of the sets G_z.*

If the theorem is false there exists for each integer n a point x_n such that no one of the sets G_z contains

$$U(x_n, 1/n).$$

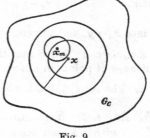

Fig. 9

Let x be a point to which some subsequence of $(x_1, x_2, \ldots)$ converges. One of the given open sets, say G_c, contains x, and therefore, for some n, the neighbourhood $U(x, 2/n)$. For at least one m greater than n, $U(x, 1/n)$ contains x_m, and therefore $U(x, 2/n)$ contains $U(x_m, 1/m)$. Thus

$$U\left(x_m, \frac{1}{m}\right) \subseteq G_c,$$

contrary to the hypothesis.

Theorem 15·3. (Heine-Borel-Lebesgue Theorem.) *A necessary and sufficient condition for a space S to be compact is that from every open covering of S a finite covering can be selected.*

Necessary. Let ϵ be the number whose existence was proved in the lemma, and A an ϵ-net in S. With each point, a, of A correlate a member, G_a, of the given covering, containing $U(a, \epsilon)$. These G_a's are finite in number (since A is a finite set), and
$$S \subseteq \bigcup_{a \in A} U(a, \epsilon) \subseteq \bigcup_{a \in A} G_a,$$
i.e. the G_a's are a finite covering of S.

Sufficient. If the infinite subset, E, of S has no limit point it is closed, and each of its points, x, has a neighbourhood U_x not meeting $E - x$. The sets $S - E$ and U_x, for all x of E, form an open covering of S from which no finite covering can be selected.

A *decreasing sequence* of sets E_1, E_2, ... is a sequence such that, for every n, $E_{n+1} \subseteq E_n$.

Theorem 15·4. *A decreasing sequence of non-null compact sets has a non-null common part.*

Let F_1, F_2, ... be the sets, and let x_n be any point of F_n. The sequence of points x_1, x_2, ... of the compact set F_1 has a sub-sequence which converges, say to x. Since x_n, x_{n+1}, x_{n+2}, ... all belong to F_n, x belongs to F_n; and therefore $x \in \bigcap F_n$.

Theorem 15·5. *If F is the common part of the sets F_n in 15·4 and ϵ is any positive number, there is an n such that every point of F_n is within ϵ of F.*

For if not, the closed, and therefore compact, subsets
$$F_n - U(F, \epsilon) \quad (= F_n \cap \mathscr{C} U(F, \epsilon))$$
of F_1 are all non-null and form a decreasing sequence, and therefore have a non-null common part. But their common part is $F - U(F, \epsilon)$, and is certainly null.

Theorem 15·4 can also be deduced, in a more general form, immediately from the Heine-Borel Theorem. A system of sets is said to be *monotone* if given any two of its members, E_1 and E_2, either $E_1 \subseteq E_2$ or $E_2 \subseteq E_1$. Any *finite* monotone system can evidently be arranged in order so that
$$E_1 \subseteq E_2 \subseteq \dots \subseteq E_k.$$

Theorem 15·6. *The common part of a monotone system of non-null compact sets, F_z, is not null.*

If F_0 is any one of the sets we may ignore all sets except those contained in F_0, for the sets containing F_0 have no effect on the common part; i.e. we may assume $F_z \subseteq F_0$. If there is no point belonging to all the sets F_z every point of F_0 belongs to at least one set $F_0 - F_z$, i.e. these sets form an open covering of F_0, regarded as a space.* Therefore a finite number of them, say

$$F_0 - F_1 \subseteq F_0 - F_2 \subseteq \ldots \subseteq F_0 - F_k,$$

cover F_0. But the union of these sets is $F_0 - F_k$, which does not contain F_0, since F_k is not null. The assumption that there is no point common to all the F_z therefore leads to a contradiction.

§ 3. INDUCED STRUCTURE. RELATIVE AND ABSOLUTE PROPERTIES

16. It was pointed out at the beginning of this chapter that any set of points, A, in a metric space, S, can itself be made into a metric space by assigning the same distance function in A as in S. The space A is then said to be *contained isometrically* in S (in contrast to "contained topologically", to be defined in Chapter III). If $E \subseteq A \subseteq S$, the derived set, closure, etc. of E may be determined regarding E either as a set of points in the space S, or as a set in the space A, and the results may be different. For example, the set $0 < \xi \leqslant \frac{1}{2}$, considered as a set in R^1, is not closed, for it does not contain its closure-point the origin; but as a set of points in $(0, 1)$ it is closed, since it contains all its closure-points *in that set*. On the other hand the question: does the sequence (x_n) converge to a? has evidently the same answer whatever set A (containing the x_n and a) is regarded as the underlying space, for the answer depends only on the distances of the x_n from a. It follows that *compactness* is a property independent of the containing space (provided the distance relations between the points of the set are undisturbed).

It appears, then, that the properties of a set of points, E, in S are of two kinds, *relative properties*, such as being open or closed, which are really *relations* between E and S, and *absolute* or

---

* Cf. the remark at the end of para. 1, and the following section.

intrinsic properties, which do not depend on the containing space, provided it induces the given metric in E. This distinction is not peculiar to the theory of sets of points, but exists in any geometrical theory in which one space can be embedded in another with the preservation of certain properties. In the differential geometry of surfaces in Euclidean space, the properties that depend only on the "first quadratic form",

$$E\,du^2 + 2F\,du\,dv + G\,dv^2,$$

are absolute properties, and can in fact be defined without reference to a surrounding space, but those that depend on the "second quadratic form" are relative to the situation of the surface in the 3-dimensional space.

Intrinsic properties of a set E can always be described by considering E itself as the containing space, and it is therefore sufficient to develop the theory of such properties for *spaces*. This procedure is implied in the definition that was given of *local compactness* (p. 43): when this definition is applied to a set of points E, "neighbourhood" and "closure" must be understood relative to the "space" E. The proof of Theorem 15·6 makes use of the intrinsic nature of compactness, F_0 being treated as a space. A more important example is the whole theory of *connection* in Chapter IV, which is greatly simplified by using the fact that connection is an intrinsic property. Of the properties already defined for sets in any metric space the following are intrinsic: being an isolated set, being bounded, separability, being dense-in-itself,* compactness, local compactness. Being dense, or nowhere dense, is a relative property.

It is useful to collect here some relations that hold between sets closed or open in a space S, and sets closed or open in a subset A of S. Relative terms, such as closed or open, E' or $\bar{E}$, when used without qualification, have reference to the space S.

Theorem 16·1. *If $E \subseteq A$, the closure of E, as a set in the space A, is $A\bar{E}$.*

For the closure of E in A is the set of points of A at distance 0 from E.

---

* The defining condition can be expressed as $E \subseteq EE'$, and EE' depends only on the distances between points of E.

Theorem 16·2. *The sets that are closed (open) in A are identical with the sets AF (AG), where F and G are respectively any closed and any open set in S.*

(i) If E is closed in A, $E = A\bar{E}$, and $\bar{E}$ is closed in S.

(ii) If $E = AF$, $\bar{E} \subseteq \bar{F} = F$, and therefore $A\bar{E} \subseteq AF = E$. Hence, by 16·1, E is closed in A.

(iii) E is open in A if, and only if, $A - E$ is closed in A, i.e. if $A - E = AF$. Since $E \subseteq A$ this is equivalent to

$$E = A - AF = A(S - F).$$

These theorems shew that the larger the set A the stronger the property of being open or closed in it; more precisely, *if $A_1 \supseteq A_2$ a subset of A_2 that is open (closed) in A_1 is open (closed) in A_2.* For if $E \subseteq A_2 \subseteq A_1$, and H is any set, $E = A_1 H$ implies $E = A_2 H$.

The following theorems go the other way:

Theorem 16·3. *If E is closed in the closed set F it is closed in S,* for $E = F\bar{E}$, the common part of two closed sets.

Theorem 16·4. *If E is open in the open set G_0 it is open in S,* for by 16·2 it is of the form $G_0 G$, the common part of two open sets.

Theorem 16·5. *If E is closed (open) in A_1 and in A_2 it is closed (open) in $A_1 \cup A_2$.*

Closed: we have

$$E = \bar{E}A_1 = \bar{E}A_2 = \bar{E}(A_1 \cup A_2),$$

the common part of $A_1 \cup A_2$ and a closed set.

Open: $E = A_1 G_1 = A_2 G_2$, where G_1 and G_2 are open. It follows that
$$E = (A_1 \cup A_2) \cap G_1 G_2.$$

For E contains both $A_1 G_1 G_2$ and $A_2 G_1 G_2$, and therefore their union; and it is contained in both $(A_1 \cup A_2) \cap G_1$ and $(A_1 \cup A_2) \cap G_2$, and therefore in their common part. Thus E is the intersection of $A_1 \cup A_2$ with an open set.

Note that in spite of the symmetry of 16·2 to 16·5, the "dual" of 16·1, "the interior of E relative to A is $A\mathscr{I}E$", is not true.

§4. COMPLETE SPACES

17. A space with metric ρ is *complete* if every fundamental sequence in it converges to a point of the space.

Examples. It was shewn above (14·4) that the cartesian spaces R^p are complete. The open interval $(0, 1)$ of R^1, considered as a space, is not complete (note following 13·1).

The Hilbert space, H^∞, is complete. Let (x_n) be a fundamental sequence, where $x_n = (\xi_{n1}, \xi_{n2}, \ldots)$. From $\lim_{m,\,n} |x_m - x_n| = 0$ it follows a fortiori that $\lim_{m,\,n} |\xi_{mr} - \xi_{nr}| = 0$ for any fixed r, and therefore that (ξ_{nr}) converges, for each fixed r, to a limit α_r. For any positive ϵ we have

$$\sum_{r=1}^{\infty} (\xi_{mr} - \xi_{nr})^2 < \epsilon$$

if m, n exceed a certain $n_0(\epsilon)$, and a fortiori

$$\sum_{r=1}^{N} (\xi_{mr} - \xi_{nr})^2 < \epsilon$$

for every N. In this finite sum we may let $m \to \infty$;

$$(1) \quad \sum_{r=1}^{N} (\xi_{nr} - \alpha_r)^2 \leqslant \epsilon$$

for all N, giving

$$(2) \quad \sum_{r=1}^{\infty} (\xi_{nr} - \alpha_r)^2 \leqslant \epsilon$$

if $n > n_0$. Let n_1 be any number exceeding n_0. Then for any N

$$\sum_{1}^{N} \alpha_r^2 \leqslant 2 \sum_{1}^{N} (\xi_{n_1 r} - \alpha_r)^2 + 2 \sum_{1}^{N} \xi_{n_1 r}^2$$

$$\leqslant 2\epsilon + 2 \sum_{1}^{\infty} \xi_{n_1 r}^2 \quad \text{by (1).}$$

Hence $\Sigma \alpha_r^2$ is convergent, i.e. $(\alpha_1, \alpha_2, \ldots)$ is a point a of H^∞.

Inequality (2) now gives $|x_n - a| \to 0$, i.e. $x_n \to a$ in H^∞.

Complete normed real or complex vector spaces are called *Banach spaces*.

Exercise. Prove that R^ω is complete.

Theorem 17·1. *In a complete space a decreasing sequence of non-null bounded closed sets, with diameter tending to zero, has a single point as intersection.*

That the intersection contains at most one point is obvious; we have to shew that it is not empty.

Let the sets be $F_1, F_2, \ldots$, and in each set F_n choose a point x_n. Given some positive ϵ, suppose $\Delta(F_{n_0}) < \epsilon$. Then $\rho(x_m, x_n) < \epsilon$ if $m, n > n_0$ and hence x_n converges, say to a. For any m, $x_n \in F_m$ if $n > m$, and therefore $a \in F_m$. Thus a is a point of $\cap F_m$.

The intersection may be empty if the condition "with diameters tending to zero" is omitted, even though the boundedness condition is retained. *Example*: take the space to be H^∞, and F_n to be $\{e_n, e_{n+1}, \ldots\}$ (notation of p. 38).

Theorem 17·2. *In a complete space the common part of an enumerable set of dense open sets is dense (and therefore not null).*

Let the open sets be $G_1, G_2, \ldots$. Let $U(x_1, \epsilon_1)$ be any spherical neighbourhood $(0 < \epsilon_1 < 1)$ and make the inductive hypothesis that x_{n-1} and ϵ_{n-1} are already determined. Since G_{n-1} is dense the open set

$$X = G_{n-1} \cap U(x_{n-1}, \tfrac{1}{2}\epsilon_{n-1})$$

is not null, and a point x_n and a positive ϵ_n, less than $1/n$, can be chosen so that $U(x_n, \epsilon_n) \subseteq X$. The closed sets $\mathscr{K} U(x_n, \tfrac{1}{2}\epsilon_n)$, a decreasing sequence with diameters tending to zero, have a common point x, which is also a point of $\cap G_n$. Since x belongs to the arbitrary neighbourhood $U(x_1, \epsilon_1)$ the theorem is proved.

Corollary 1. *A complete space is not the union of an enumerable set of nowhere dense closed subsets.* This follows from 17·2 on taking complements. The example of Q^1, the space of rational numbers, whose individual points are closed nowhere dense subsets, shews that this corollary may fail in a non-complete space.

The following generalisation of Corollary 1 is known as "Baire's Density Theorem".

Corollary 2. *If the set E in a complete space is the union of an enumerable set of nowhere-dense sets, $S - E$ is dense.* If $E_1, E_2, \ldots$ are the nowhere-dense sets, the sets $S - \bar{E}_n$ are (by definition) dense in S. Hence $\cap(S - \bar{E}_n) = S - \cup \bar{E}_n$ is dense, and a fortiori $S - \cup E_n = S - E$ is dense.

Example. A space is *semi-compact* if it is the union of an enumerable set of compact sets. Since $R^p = \cup \overline{U(o,n)}$, it is semi-compact; but H^∞ and R^ω are not. This follows from Corollary 1 of 17·2, since all compact sets in H^∞ and R^ω are nowhere-dense (Example 5, p. 44).

Theorem 17·3. *In a complete space the existence of an ϵ-net for every positive ϵ is a sufficient condition for compactness.* (That the condition is necessary, in any space, was shewn in 15·1.)

Let E_0 be any infinite subset, and suppose, inductively, that an infinite subset E_{n-1} of E_0 has been determined. Since S is covered by a finite set of neighbourhoods $U(a, 1/n)$, one such neighbourhood, say $U(a_n, 1/n)$ must contain an infinite subset E_n of E_{n-1}. The closed sets

$$F_n = \bigcap_{r=1}^{n} \overline{U(a_r, 1/r)}$$

form a decreasing sequence with diameters tending to zero, and they are non-null since $E_n \subseteq E_r \subseteq U(a_r, 1/r)$ for $r \leqslant n$. Let x be a point of $\cap F_n$. Then $x \subseteq \overline{U(a_n, 1/n)}$ and therefore

$$E_n \subseteq U(a_n, 1/n) \subseteq U(x, 3/n).$$

Since E_n is infinite, x is a limit-point of E_0.

Exercise. Use this theorem to prove that I^ω is compact.

18. The following Completion Theorem shews that incomplete metric "spaces" are most naturally regarded as unclosed sets of points in a complete space. A complete space S cannot be contained isometrically† as an unclosed set in any other space T; for if (x_n) of S converges to a in T, (x_n) is a fundamental sequence and therefore has a limit b in S. Hence $b = a$, i.e. $a \in S$.

Theorem 18·1. *Every metric space M is contained isometrically as a dense subset in a complete metric space M^*.*

The proof is a generalisation of the process of constructing the real numbers by filling "holes" in the set of rational numbers.

Let two fundamental sequences, (x_n) and (y_n) in M be called "equivalent" if $\lim \rho(x_n, y_n) = 0$. It is easily seen that this is indeed an equivalence relation, and that a class of equivalent sequences either all converge to the same point of M, or do not converge at all. In the second case the class of sequences is said

† See beginning of para. 16.

to determine a *hole*. With every hole we associate a new "point", the new points associated with different holes being distinct from each other and from all points of M (the "old" points). Let M^* be the set of old and new points, and let us denote an old point by x, a new one by y, and one that may be either old or new by z.

Let z_1 and z_2 be two points of M^*, and let (x_{in}) be, for $i = 1$ and 2, a sequence of old points *associated* with z_i, i.e. a sequence converging to z_i if it is an old point, a fundamental sequence of the corresponding hole if z_i is new.

The sequence of numbers $\rho(x_{1n}, x_{2n})$ converges to a limit as $n \to \infty$, for

$$
\begin{aligned}
| \rho(x_{1m}, x_{2m}) - \rho(x_{1n}, x_{2n}) | &\leqslant | \rho(x_{1m}, x_{2m}) - \rho(x_{1m}, x_{2n}) | \\
&\quad + | \rho(x_{1m}, x_{2n}) - \rho(x_{1n}, x_{2n}) | \\
&\leqslant \rho(x_{2m}, x_{2n}) + \rho(x_{1m}, x_{1n}) \to 0,
\end{aligned}
$$

as m, $n \to \infty$. This limit is not altered by replacing (x_{1n}) or (x_{2n}) by any equivalent sequence, for if, say, (x_{3n}) is equivalent to (x_{1n}),

$$
| \rho(x_{1n}, x_{2n}) - \rho(x_{3n}, x_{2n}) | \leqslant \rho(x_{1n}, x_{3n}) \to 0.
$$

The limit is therefore a function of the points z_1 and z_2, and may be denoted by $\sigma(z_1, z_2)$.

σ *satisfies the conditions* m_1, m_2, m_3 *for a metric* (p. 17). If $\sigma(z_1, z_2) = 0$, $\lim \rho(x_{1n}, x_{2n}) = 0$, and therefore the two sequences either converge to the same point of M or belong to the same hole: m_1 is satisfied. The other two conditions are easily verified by proceeding to the limit in the corresponding conditions for the space M. If x_1 and x_2 are old points $\sigma(x_1, x_2) = \rho(x_1, x_2)$, since

$$
\rho(x_{1n}, x_{2n}) \to \rho(x_1, x_2).
$$

It has thus been shewn that σ determines a metric space M^* in which M is contained isometrically.

If (x_n) *is a fundamental sequence of old points defining the new point* y, *then* $x_n \to y$ *in* M^*, for since $(x_n, x_n, x_n, \ldots)$ is one of the sequences converging to x_n in M,

$$
\sigma(x_n, y) = \lim_{m \to \infty} \rho(x_n, x_m) \underset{n}{\to} 0,
$$

since (x_n) is a fundamental sequence. Thus M *is dense in* M^*.

Let (z_n) be any fundamental sequence in M^*, and x_n an old point such that $\sigma(z_n, x_n) < 1/n$. Then

$$
\begin{aligned}
\rho(x_m, x_n) &= \sigma(x_m, x_n) \\
&\leqslant \sigma(x_m, z_m) + \sigma(z_m, z_n) + \sigma(z_n, x_n) \\
&\to 0
\end{aligned}
$$

as $m, n \to \infty$. Thus (x_n) is a fundamental sequence in M. Let z be the corresponding point of M^*. Then

$$
\sigma(z_n, z) \leqslant \sigma(z_n, x_n) + \sigma(x_n, z) \to 0,
$$

since $x_n \to z$ in M^*. Thus the sequence (z_n) is convergent in the space M^*, which is therefore complete.

Chapter III

HOMEOMORPHISM AND CONTINUOUS MAPPINGS

§ 1. EQUIVALENT METRICS. TOPOLOGICAL SPACES. HOMEOMORPHISM

1. If the definitions given in Chapter II are now reconsidered it will be found that most of them make no direct reference to the underlying metric, but are expressed in terms of a small number of fundamental concepts introduced early in the chapter—namely closure, open set and closed set, and convergent sequence. It is with properties expressible in these terms that topology is concerned.

In any given set of things there are numerous distance functions satisfying the conditions m_1, m_2, m_3, and any two of them determine (by definition) different *metric spaces*. But it may happen that if the definitions of the four fundamental concepts just mentioned are interpreted first in terms of one metric, ρ, and then in terms of another, σ, the results are precisely the same; i.e. the *closure* $\bar{E}$ of every set E is the same whether ρ or σ is used in determining it, the same sets are *open* and *closed* by either standard, and (x_n) *converges to a* by ρ-standards if, and only if, it converges to a by σ-standards. These relations between the metrics are not independent, as the following theorem shews.

Theorem 1·1. *Let S and T be metric spaces with the same points, but different metrics. If any one of the following statements is true, all are true (the suffix S meaning "in space S"):*

(A) *the closed sets of S are identical with those of T;*

(B) *the open sets of S are identical with those of T;*

(C) *for all a, $x_n \underset{S}{\rightarrow} a$ if, and only if, $x_n \underset{T}{\rightarrow} a$;*

(D) *for all sets E, $\bar{E}_S = \bar{E}_T$.*

(A) implies (B) by taking complements. (B) implies (C) since $x_n \rightarrow a$ if, and only if, every open set containing a contains x_n for all but a finite number of n. (C) implies (D) since $\bar{E}$ is the set of limits of convergent sequences in E. (D) implies (A) because E is closed if, and only if, $E = \bar{E}$.

Two metrics in the same set are *topologically equivalent* if one (and therefore all) of conditions (A)–(D) is satisfied. A more convenient criterion than any of these is

Theorem 1·2. *A necessary and sufficient condition that the metrics ρ and σ be topologically equivalent is that for each fixed a, $\rho(a,x)$ and $\sigma(a,x)$ tend to zero together; i.e. given any positive ϵ there exists a positive δ such that $\rho(a,x) < \delta$ implies $\sigma(a,x) < \epsilon$, and $\sigma(a,x) < \delta$ implies $\rho(a,x) < \epsilon$.*

The condition is necessary, for if, for example, the first half is not satisfied, then for some point a and a certain positive ϵ there is, for each n, a point x_n such that $\rho(a,x_n) < n^{-1}$ but $\sigma(a,x_n) > \epsilon$. Thus $x_n \to a$ by ρ-, but not by σ-standards.

The condition is sufficient, for if it is satisfied, and if $x_n \to a$ by ρ-standards, i.e. if

$$\lim_{n \to \infty} \rho(x_n, a) = 0,$$

it follows that

$$\lim_{n \to \infty} \sigma(x_n, a) = 0,$$

i.e. that $x_n \to a$ by σ-standards. Thus ρ-convergence and σ-convergence are the same, and it follows that ρ and σ are completely equivalent.

Corollary. *A necessary and sufficient condition for ρ and σ to be equivalent is that, for each a, $\sigma(a,x)$ is continuous at a by ρ-standards, and $\rho(a,x)$ by σ-standards.*

The geometrical meaning of these criteria is that points which are near each other according to the one metric are also near each other according to the other, or more concretely, the change of metric represents a bending and stretching of the space without tearing.

Examples. 1. The function

$$| \xi_1 - \eta_1 | + | \xi_2 - \eta_2 |$$

is a metric in R^2 equivalent to $| x - y |$.

2. If the open interval $(0, 1)$ is regarded as a space, the function

$$\rho_0(\xi, \eta) = | \xi^{-1} - \eta^{-1} |$$

defines a metric equivalent in it to $| \xi - \eta |$.

N T

3. The following are examples of metrics in the set of real numbers *not* equivalent to $|\xi - \eta|$:

(*a*) $\theta_1(\xi, \eta) = 1$ for every distinct pair, ξ and η.

(*b*) $\theta_2(\xi, \eta) = |\xi - \eta| + 1$ if one of the numbers ξ, η is positive and the other is not, $\theta_2(\xi, \eta) = |\xi - \eta|$ in all other cases. The sequence $1, \frac{1}{2}, \frac{1}{3}, \ldots$ converges to o according to the metric $|\xi - \eta|$ but not according to θ_2. (Verify that the conditions m_i are satisfied!)

4. No *norm* in the space of sequences of real numbers can determine a metric topologically equivalent to that of R^ω. For if $|x - y|$ is the original R^ω-metric and $|\|x\||$ any norm, and if e_r is as on p. 38, then for all values of (λ_n), $\lambda_n e_n \to o$ in R^ω, since $|\lambda_n e_n - o| \leqslant 1/n^2$. But if $\lambda_n = 1/|\|e_n\||$, $|\|\lambda_n e_n - o\|| = 1$ and $\lambda_n e_n \not\to o$ by the $|\|x - y\||$-metric.

Equivalent bounded metric. *If $\rho(x, y)$ is a metric in a space S, $\rho^*(x, y) = \min(1, \rho(x, y))$ is an equivalent metric.* The conditions m_1 and m_2 for a metric are clearly satisfied, and m_3 follows from

$$\min(\alpha + \beta, \gamma) \leqslant \min(\alpha, \gamma) + \min(\beta, \gamma).$$

Cf. footnote, p. 20. The metrics are equivalent, for $\rho(a, x) < \epsilon$ implies $\rho^*(a, x) < \epsilon$; and if $\rho^*(a, x) < \min(1, \epsilon)$ then

$$\rho(a, x) = \rho^*(a, x) < \epsilon.$$

Thus to *any metric corresponds a topologically equivalent bounded metric*.

Exercise. If ρ is any metric in S,

$$\sigma(x, y) = \frac{\rho(x, y)}{1 + \rho(x, y)}$$

is an equivalent bounded metric.

2. A property of metric spaces, or a relation between sets in a metric space, is said to be *topological* (or topologically invariant) if it is unaffected by a change from one metric to another topologically equivalent one. All properties and relations expressible in terms of the four fundamental concepts are therefore topologically invariant, for example, compactness, local compactness, separability, and being dense-in-itself. On the other hand, boundedness is not topologically invariant (as the preceding Exercise shews), nor is completeness, as will be shewn below. Thus the Weierstrass Theorem is not a theorem of topology.

It is clear that the metric is not part of the subject-matter of topology, but only an accessory used in the definitions of certain concepts, and it is natural to say that a class of equivalent metrics in a set S are merely different ways of specifying the same *metrisable topological space*, or the same *topological structure* in S. Any metric space with metric ρ determines a metrisable topological space, having ρ as one of its admissible metrics.

The significance of the qualifying word "metrisable" is that other topological spaces can be considered, in which the closed and open sets are not specified by first setting up a metric, but in some other way. It will be clear from what has been said above that the specification of the open sets of a space, in whatever manner, is a sufficient basis for interpreting all our topological definitions given in Chapter II, though it is not to be assumed that all the theorems of that Chapter will remain valid without some restriction in the choice of the "open" sets. It may happen that a topological space specified directly in this way may nevertheless admit a metric, i.e. that a metric ρ can be set up in it so that the open sets defined in terms of ρ, as in Chapter II, are identical with those originally specified. Such a topological space is *metrisable*.[10]

It usually leads to no confusion to use the same symbol in particular instances for a metric space and the metrisable topological space it determines. The names R^p, H^∞, R^ω, I^ω will be used in this way.

Examples. 1. Take as points all sequences of real numbers, and call any set of points satisfying a finite set of inequalities

$$\alpha_r < \xi_r < \beta_r \quad (r = 1, 2, \ldots, k)$$

a *slab*. Let the "open sets" be, by definition, all unions of slabs. *This topological space is R^ω*, i.e. it is metrisable by the metric defined on p. 19, which we will call ρ. Let us call unions of slabs "σ-open", and the open sets determined by the metric ρ "ρ-open". (1) Slabs are ρ-open (Chapter II, para. 3, Exercise 2 and Theorem 3·7) and therefore unions of them are ρ-open. (2) Let E be any ρ-open set, and suppose that $U_\rho(a, \epsilon) \subseteq E$, and that

$$\sum_{r_0}^{\infty} \frac{1}{r^2} < \tfrac{1}{2}\epsilon.$$

Let A_a be the slab

$$\alpha_r - \frac{\epsilon}{2r_0} < \xi_r < \alpha_r + \frac{\epsilon}{2r_0} \quad (r = 1, 2, ..., r_0),$$

where a is $(\alpha_1, \alpha_2, ...)$. Then if $x \in A_a$

$$\rho(x, a) \leqslant \sum_1^{r_0-1} \frac{1}{r_2} \frac{\epsilon}{2r_0} + \sum_{r_0}^{\infty} \frac{1}{r^2} < \epsilon,$$

i.e. $A_a \subseteq U_\rho(a, \epsilon)$. If a slab A_z contained in E is similarly associated with each point z of E, $\bigcup_z A_z = E$, and hence E is σ-open. Thus the two spaces have the same open sets, and are therefore identical.

It has thus been shewn that the "slabs" are a base in R^ω.

2. The (metric) product-space $S \times T$ of two metric spaces with metrics ρ and σ was defined in Chapter II, 1, Example. If ρ' and σ' are topologically equivalent metrics in S and T respectively, it is easily seen by 1·2, that the metric $\rho' + \sigma'$ in $S \times T$ is topologically equivalent to $\rho + \sigma$. Thus all such metric product-spaces define a *topological product-space*, which will also be denoted by $S \times T$.

From Example 1, para. 1, it follows that the topological spaces $R^1 \times R^1$ and R^2 are identical.

Exercise. Define the topological product-space $S_1 \times S_2 \times ... \times S_k$, and prove that the topological spaces $R^1 \times R^1 \times ... \times R^1$ (p factors) and R^p are identical.

If $A \subseteq S$, topologically equivalent metrics in S determine topologically equivalent metrics in A, for the closed sets of A are the sets AF, where the F are the closed sets of S, and are therefore the same for the two metrics. The corresponding topological structure in A is said to be induced by that in S, and A is a *subspace* of S. It may happen that the topological space A has, besides the S-metrics, others that are not capable of extension to the whole space S.

No metric defined in the topological space R^1 agrees with the metric ρ_0 (Example 2, para. 1) in $(0, 1)$. For such a metric, being defined over the compact set $< 0, 1 >$, would be bounded there, and a fortiori in $(0, 1)$. Thus a set E in a space S may admit equivalent metrics which cannot be extended to a metric valid in the whole of S.

3. Two metrisable topological spaces S and T are *homeomorphic* if there exists a $(1, 1)$-correlation between them, which correlates with each closed set in S a closed set in T, and *vice versa*. (The

Homeomorphism is an equivalence relation, and a space and its homeomorphs have all their topological properties in common.

The definition may be extended to homeomorphisms between sets of points, E_1 and E_2, in the same or different spaces S and T, by regarding E_1 and E_2 as spaces with the topologies induced by S and T respectively. The term "closed set" is then to be understood relative to E_1 and E_2 respectively: it does not follow from the fact that E_1 and E_2 are homeomorphic that if E_1 is closed in S then E_2 is closed in T (see Example 2 below).

Theorem 3·1. *If a* (1, 1)-*correlation is given between the metrisable topological spaces* S *and* T, *each of the following statements implies all the others:*

(A) *each closed set in* S *is correlated with a closed set in* T *and vice versa;*

(B) *each open set in* S *is correlated with an open set in* T *and vice versa;*

(C) $x_n \to a$ *in* S *if, and only if,* $x'_n \to a'$ *in* T *(dashes denoting correlates);*

(D) *the closures of correlated sets are themselves correlated sets.*

Hence each of the four is a necessary and sufficient condition for the correlation to be a homeomorphism.

The proof is similar to that of 1·1.

Theorem 3·2. *If the topological spaces* S *and* T *admit the metrics* ρ *and* σ, *a necessary and sufficient condition for the* (1, 1)-*mapping* f *of* S *on to* T *to be a homeomorphism is that for each point* a *of* S, $\rho(a, x)$ *and* $\sigma(fa, fx)$ *tend to zero together.*

This condition is to be understood in the sense of 1·2, and 3·2 follows from (C) of 3·1, as 1·2 follows from (C) of 1·1.

Corollary. *The function* $\sigma(fx, fy)$ *is a metric in* S *topologically equivalent to* ρ.

An immediate consequence of 3·2 is the following criterion for a homeomorphism between sets in R^p: *if* f *is a* (1, 1)-*mapping of the subset* E_1 *of* R^p *on to* E_2 *of* R^q,* *and if, for each* x *of* E_1 *and each* y *of* E_2 *the coordinates of* $f(x)$ *and* $f^{-1}(y)$ *are continuous functions of those of* x *and* y *respectively, then* f *is a topological mapping.*

* The notation R^p implies that p is finite: R^ω is *not* included.

Examples. 1. If $\alpha < \beta$ the interval $<0, 1>$ is mapped topologically on to $<\alpha, \beta>$ by the function $\phi(\xi), = \alpha + \xi(\beta - \alpha)$; for

$$|\phi(\xi) - \phi(\eta)| = (\beta - \alpha)|\xi - \eta|.$$

2. The open interval $(0, 1)$ and the line R^1 are homeomorphic. Mapping function:

$$\frac{1}{1 - \xi} - \frac{1}{\xi}.$$

This function also maps $(\frac{1}{2}, 1)$ topologically on to $(0, \infty)$, and $<\frac{1}{2}, 1)$ on to $<0, \infty)$. Combining the results of 1 and 2 we see that all the open, half-open, and closed intervals of R^1 fall into the following three classes, sets in each row being homeomorphic to each other:

(a) R^1, (α, β), $(-\infty, \alpha)$, (α, ∞);
(b) $(\alpha, \beta>$, $<\alpha, \beta)$, $(-\infty, \alpha>$, $<\alpha, \infty)$;
(c) $<\alpha, \beta>$.

It will be shown in Chapter IV that sets in different rows are not homeomorphic. (Clearly (c) is not the homeomorph of any of the rest, because (c) is compact and the other sets are not.)

3. The set $[|\xi_r| < 1$, all coordinates$]$

in R^p or R^ω is the homeomorph of the whole space. (From the previous example and that on p. 38.)

4. The sets

$$\left[\sum_1^p \xi_r^2 \leqslant 1\right] \quad \text{and} \quad [|\xi_r| \leqslant 1,\ r = 1, 2, ..., p]$$

in R^p are homeomorphic. Each radius of the "sphere" is mapped linearly on the radius of the "cube" in the same direction; i.e. o is mapped on itself, and if x is any other point of the spherical region, and ox cuts the sphere in y and the cube (the frontier of the second set) in z, $f(x)$ is the point, t, of oz such that $ox/oy = ot/oz$. It is easily proved, by 3·2, that this is a topological mapping.

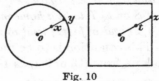

Fig. 10

This is a special case of the theorem that *if G_1 and G_2 are bounded open convex sets in R^p, $\bar{G}_1$ and $\bar{G}_2$ are homeomorphic.*

5. Every topological mapping, ϕ, of $<0, 1>$ on to itself is a monotone function of ξ. For if not there are three points of the interval such that $\xi_1 < \xi_2 < \xi_3$, but $\phi(\xi_2)$ is not in the interval $<\phi(\xi_1),\ \phi(\xi_3)>$. Suppose, for example, that $\phi(\xi_2) < \phi(\xi_1) < \phi(\xi_3)$. Then since ϕ is continuous it takes the value $\phi(\xi_1)$ for a value of ξ between ξ_2 and ξ_3; i.e. ϕ is not $(1, 1)$, contrary to the assumption.

It follows that the end-points are mapped either on themselves, in which case ϕ is "sense-preserving"; or on each other, and ϕ is "sense-reversing".

6. The set $|\xi_r| \leqslant 1$ (all coordinates) in R^ω is the homeomorph of I^ω. [Use Exercise 1 on p. 39. It also follows that the set $|\xi_r| < 1$ in R^ω is the homeomorph of $|\xi_r| < 1/r$ in H^∞, and hence, using Example 3 above, the question whether R^ω and H^∞ are homeomorphic is reduced to the question: is the set $|\xi_r| < 1/r$ in H^∞ the homeomorph of the whole space H^∞? The proof in Example 4, p. 58, that no norm in the space whose points are all sequences of real numbers defines a metric in R^ω, shews that there is no *linear* homeomorphism, f, of R^ω on to any normed real vector space, for then $\||f(x)|\|$ would be a topologically equivalent norm in R^ω. This does not settle the question of the existence of other homeomorphisms, the answer to which appears to be unknown at present.]

7. A set $[k]$ ("k-flat") in any real normed vector space is the homeomorph of R^k. [Use the lemma proved in the Example, p. 44.]

8. If $\phi(\xi_2, \xi_3, ..., \xi_p)$ is a real continuous function in any subset E of $[\xi_1 = 0]$ in R^p, the graph $[\xi_1 = \phi(x)]$ of ϕ, for

$$x = (\xi_2, \xi_3, ..., \xi_p) \in E,$$

is homeomorphic with E, under the correlation $x \to (\phi(x), x)$. For if x and y are points of E,

$$|x - y| \leqslant |(\phi(x), x) - (\phi(y), y)| \leqslant |x - y| + |\phi(x) - \phi(y)|.$$

More generally, if ϕ is a real continuous function in any space S, the set $[\xi_1 = \phi(x)]$ in $R^1 \times S$ is the homeomorph of S.

4. In the preceding examples certain pairs of spaces are shewn to be homeomorphic. To prove that two spaces are *not* homeomorphic, a topologically invariant property must be found that belongs to one but not the other. For example, no *compact* space (segment, circle, n-sphere, I^ω) can be the homeomorph of a space that is not compact (line, R^p, open set in R^p, H^∞). Since R^p is *locally compact* and H^∞ is not, these spaces are not homeomorphic.

Such crude methods do not carry us far in discriminating between spaces: even the proof that R^p and R^q are not homeomorphic if $p \neq q$ will be found in Chapter v to require the use of powerful separation theorems.

Example. Let A be the set of all points $(\xi_1, \xi_2, ..., \xi_k, 0, 0, ...)$ of H^∞ with only a finite number of non-zero coordinates. This set, being the union of homeomorphs of R_1, R_2, R_3, ... (Example 7 above), is semi-compact (Example, p. 53). Semi-compactness is clearly a topological property, which is not possessed by H^∞ (same example). Hence A and H^∞ are not homeomorphic. On the other hand A is not locally compact (since it contains the points ϵe_1, ϵe_2, ..., p. 38) and so is not the homeomorph of any R^p. Thus A is a normed vector space intermediate in character between R^p and H^∞.

5. A complete metric space cannot (as we have seen in II. 18) be embedded isometrically as a dense set in any larger space; but the topological space that it determines may well be capable of such enlargement, with the help of a suitable change of metric. An important example is the extension of the "open" cartesian space R^p, by the addition of one point, to form the "closed" cartesian space Z^p. (The traditional names "open" and "closed" for these spaces are only loosely connected with open and closed sets, and should be regarded as mere labels for the spaces.)

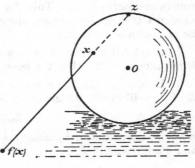

Fig. 11

Consider first the case $p = 2$. Let S^2 be the unit sphere, $[\xi_1^2 + \xi_2^2 + \xi_3^2 = 1]$, in R^3, and let z be the point $(0, 0, 1)$. A $(1, 1)$-correlation is set up between the set $S^2 - (z)$ and the plane $[\xi_3 = -1]$ in R^3, by correlating each point x of $S^2 - (z)$ with the point x' in which the ray zx cuts the plane. The coordinates (ξ_r) of x and (ξ_r') of x' are related by

$$\xi_1' = 2\xi_1/(1-\xi_3), \quad \xi_2' = 2\xi_2/(1-\xi_3), \quad \xi_3' = -1.$$

The first two equations, combined with $\xi_1^2 + \xi_2^2 + \xi_3^2 = 1$ are uniquely soluble as equations for ξ_1, ξ_2, ξ_3, with the solution

$$\xi_1 = \frac{4\xi_1'}{D+4}, \quad \xi_2 = \frac{4\xi_2'}{D+4}, \quad \xi_3 = \frac{D-4}{D+4}, \quad D = \xi_1'^2 + \xi_2'^2.$$

This confirms the geometrically obvious fact that the correlation between the plane and $S^2 - (z)$ is $(1, 1)$, and from the continuity of all the functions (subject to $\xi_3 \neq 1$) the correlation is a homeomorphism. It has thus been shewn that the "punctured sphere" is the homeomorph, under the stereographic projection, of the plane [$\xi_3 = -1$], and therefore of R^2 under, say, the mapping $f: S^2 - (z) \to R^2$. Hence if σ is a metric in S^2, $\rho(x, y) = \sigma(fx, fy)$ is a metric in R^2 topologically equivalent to $|x-y|$. (3·2, Corollary.)

Let a new topological space Z^2 be formed by taking as points the set $R^2 \cup (w)$ where w is a new point, and as an admissible metric the function ρ, just defined, extended to include w by defining $f(z)$ to be w. Then S^2 mapped (1, 1) and isometrically (relative to ρ) on to Z^2 by f, and therefore Z^2 is the homeomorph of S^2 and hence a compact space. In R^2 the metric ρ is equivalent to $|x-y|$, and therefore Z^2 is a *topological extension* of R^2, that is, the topology which Z^2 induces in its subset R^2 is identical with the ordinary topology of R^2 determined by the metric $|x-y|$.

If (z_n) is a sequence of points in R^2, the condition that $z_n \to w$ in Z^2 is that $|z_n| \to \infty$. For this reason w is called the "point at infinity". Since z is in no way distinguished from other points of the sphere, the point at infinity is not topologically different from the other points of Z^2: the closed plane is a "homogeneous" space. This mapping will frequently be used later, and the reader is advised to make himself familiar with its properties. Straight lines in the plane project into circles through w, circles in the plane into circles not through w.

The "closed cartesian space" Z^p is defined analogously to Z^2, as a compact, metrisable, topological space, homeomorphic with S^p, by adding a point w to R^p, and using the corresponding homeomorphism

$$\xi_r' = 2\xi_r/(1 - \xi_{p+1}) \quad (r = 1, 2, \ldots, p),$$

of $S^p - (z)$ on to $\xi'_{p+1} = -1$ to define an equivalent metric that can be extended to w.

From this point on "space (with metric ρ)" means "metrisable topological space (admitting ρ as one of its metrics)".

§ 2. CONTINUOUS MAPPINGS

*6. Let A and B be any two sets of things. *A function in A with values in B*, or *a transformation of A into B* is determined by correlating with each element, x, of A an element, $f(x)$, of B, which is called the f-image of x. If every element of B is the f-image of at least one element of A, f is said to be a transformation of A *on to B*.

Any suffix notation, c_x, is an alternative notation for a transformation of the set of suffixes into the set to which the elements c belong; for the symbols c_x and $c(x)$ are evidently essentially the same.

Examples. 1. Any real function, $f(\xi)$, of a real variable defined for every ξ is a transformation of R^1 into itself. If f takes all real values it is a transformation of R^1 on to itself.

2. If A is any set of points in the plane R^2, a transformation ("parallel projection") of A into the ξ_1-axis is determined by the rule $f(\xi_1, \xi_2) = (\xi_1, 0)$.

If E is any subset of A we denote by $f(E)$ the set of all image-points of E, i.e.

$$f(E) = \bigcup_{x \in E} f(x).$$

The condition that f be a transformation of A *on to B* is therefore that $B = f(A)$.

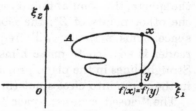

Fig. 12.   Example 2

7. A transformation, f, of a space† S into a space T is *continuous at a* if $x_n \to a$ in S implies $f(x_n) \to f(a)$ in T.‡ A *continuous mapping*, or simply a *mapping*§ of S into T is a transformation that is continuous at every point. If f maps S continuously *on to T*, T is the *continuous image* of S. These definitions apply without modification to mappings of sets of points.

---

* This paragraph belongs logically to Chapter I.

† See the end of para. 5.

‡ An alternative form of the definition, suitable for general topological spaces, is given in para. 9.

§ Outside topology "mapping" is often used in the sense of "transformation". In this book it will be used only in the sense of "continuous mapping", though the word "continuous" is sometimes retained for emphasis.

When S is R^1, the set of real numbers, f is a real function, and the definition of continuity agrees with that already given in Chapter II (para. 3).

Theorem 7·1. *A necessary and sufficient condition for the transformation f of S, (metric ρ), into T, (metric σ), to be continuous at a is that for each positive ϵ there exist a positive δ such that $\rho(x, a) < \delta$ implies $\sigma(fx, fa) < \epsilon$.*
The proof is exactly like that of 1·1.

The simpler properties of mappings are well illustrated by the *projections* of sets of points in R^p on to various linear subspaces. For example, if E is any set of points in R^2 the "parallel projection", p, of E on to the ξ_1-axis (cf. Example 2, previous page) is continuous. For

$$\rho(px, py) = |\xi_1 - \eta_1| \leqslant \rho(x, y).$$

The number of points in $p^{-1}(z)$, the set of points mapped on to z, may be 0, or any finite number, or infinite. If E is the unit circle the number is 0, 1 or 2; if E is the interior of the unit circle the number is 0 or c. Conical projections from a point on to a plane in R^3 are also continuous, and the extreme case of the mapping of the whole space on to a single point (which is also continuous) may be regarded as an "inward" projection.

Exercise. Prove that the Cantor set (II. 11) can be mapped continuously on to the segment $< 0, 1 >$. [In the ternary fraction (without 1's) for x replace all 2's by 1's and interpret in the scale of 2: this is $f(x)$.]

8. *Uniformity of continuity.* **Theorem 8·1.** *Let f be a mapping of a compact space A, (metric ρ), into a space B, (metric σ). Then, given any positive ϵ, there exists a positive δ such that*

$$\rho(x, y) < \delta \quad implies \quad \sigma(fx, fy) < \epsilon.$$

Let a positive number $\delta(a)$ be associated with each point a of A, in such a way that

$$\rho(x, a) < 2\delta(a) \quad implies \quad \sigma(fx, fa) < \tfrac{1}{2}\epsilon.$$

From the set of neighbourhoods $U(x, \delta(x))$—an open covering of A—let a finite covering

$$U(a_r, \delta_r) \quad (r = 1, 2, \ldots, k; \; \delta_r = \delta(a_r)),$$

be selected, and let δ be the least of the numbers δ_r.

If $\rho(x, y) < \delta$, then $x \in U(a_r, \delta_r)$ for some r, and

$$\rho(a_r, y) \leqslant \rho(a_r, x) + \rho(x, y) < 2\delta_r.$$

Hence $\qquad\qquad \sigma(fx, fa_r) < \tfrac{1}{2}\epsilon, \quad \sigma(fy, fa_r) < \tfrac{1}{2}\epsilon,$

and therefore $\qquad\qquad \sigma(fx, fy) < \epsilon.$

9. *Properties invariant under a mapping.* If there exists a *topological* mapping of a space S on to the space T all the topological properties of the two spaces are identical. If it is known only that a *continuous* mapping of S on to T exists it is naturally not possible to infer as much, and the spaces may in fact be very different. A striking example is that the dimension of a space may be lowered or raised to any extent in a continuous mapping. (See Exercise, para. 7, and Chapter IV, para. 9.) There are, however, several important properties which are preserved under any continuous mapping, notably compactness and (as will be shewn in the next chapter) connectedness.

Theorem 9·1. *The continuous image of a compact set E is compact.*

If $y_1, y_2, \ldots$ is any sequence of points in the image set, and $x_1, x_2, \ldots$ any sequence of points such that $f(x_n) = y_n$, there exists a subsequence, $x_{n_1}, x_{n_2}, \ldots$, of the x_n converging to a point x of E. Since f is continuous the sequence $f(x_{n_r})$ converges to $f(x)$. Thus $y_1, y_2, \ldots$ has a subsequence converging to a point of $f(E)$.

Corollary. *If f maps a compact space S into a space T, it maps closed sets in S on to closed sets in T.* For if F is closed in S it is compact. Therefore $f(F)$ is compact, and therefore closed in T.

If S is not compact, then even if f maps S continuously on the whole of T it is not necessarily true that all closed sets in S are mapped on closed sets in T. *Example*: The set E_1 consisting of the points $0, 1, 2, \ldots$ of R^1 is mapped continuously on to

E_2, consisting of the points $0, 1, \frac{1}{2}, \frac{1}{3}, \ldots$ of the same space, by mapping o on itself and n on $1/n$. (All mappings of an isolated space are continuous.) But the subset $E_1 - (o)$ is closed in E_1, its image $E_2 - (o)$ is not closed in E_2. It is, however, always true that the *inverse image* (in a sense now to be defined) of a closed set in T is closed in S.

If E is any set in T, $f^{-1}(E)$ denotes the set of all points of S that are mapped on points of E. (Thus in Example 2 on p. 66, if x is a point of the ξ_1-axis, $f^{-1}(x)$ is the set of points of A vertically above x.) From the assumption that $f(S)$ is the whole of T it follows that $f^{-1}(T - E) = S - f^{-1}(E)$, for a necessary and sufficient condition that x belong to $S - f^{-1}(E)$ is that $f(x)$ is not in E, i.e. that $f(x)$ is in $T - E$. Hence if the sets E_1 and E_2 in T do not meet, $f^{-1}(E_1)$ and $f^{-1}(E_2)$ do not meet. (No such results hold, in general, for the mapping f itself.)

Theorem 9·2. *A necessary and sufficient condition for a transformation f of S on to T to be a continuous mapping is that, for all closed sets F of T, $f^{-1}(F)$ is closed in S; and "closed" may be replaced throughout by "open".*

Necessary. If the sequence of points $x_1, x_2, \ldots$ of points of $f^{-1}(F)$ converges to x, the image-points $f(x_n)$ converge to $f(x)$. Since $f(x_n) \in F$ for all n, $f(x) \in F$, and hence $x \in f^{-1}(F)$. Thus $f^{-1}(F)$ is closed. If G is open,

$$(1) \qquad f^{-1}(G) = S - f^{-1}(T - G),$$

an open set, as has just been proved.

Sufficient. First suppose the open-set condition is satisfied. Let a be any point of S, and ϵ any positive number. The set of points mapped on to the open set $U_\sigma(fa, \epsilon)$ in T is an open set containing a, and therefore contains a neighbourhood $\bar{U}_\rho(a, \delta)$. Therefore, by 7·1, f is continuous at a. The closed-set condition implies the open-set condition, by equation (1).

These necessary and sufficient conditions provide alternative forms of the definition of a continuous mapping, suitable for use in general topological spaces.

It is clear from Theorem 3·1 that *a necessary and sufficient condition that a* (1, 1)-*transformation, f, of S on to T be a homeomorphism is that both f and f^{-1} be continuous.*

It is not in general true that, if f is a (1, 1) and continuous mapping of S on T, f is topological, i.e. that f^{-1} is also continuous.

Example. Let S be the subspace $<0, 2\pi)$ of R^1, and T the unit circle in R^2. Then $f(\xi) = (\cos\xi, \sin\xi)$ is a (1, 1) continuous mapping of S on to T; but f^{-1} is not (by 9·2) continuous, since $<\pi, 2\pi)$ is closed in S, but $f(<\pi, 2\pi))$ is not closed on the circle.

In compact spaces, however, the continuity of f alone is sufficient:

Theorem 9·3. *If a continuous mapping of a compact space S on to a space T is* (1, 1), *it is topological.*

Let f be the mapping, and F any closed set in S. By 9·1, Corollary, $f(F)$ is closed in T, and therefore by 9·2, f^{-1} is continuous.

Chapter IV

CONNECTION

§ 1. CONNECTED SETS

1. Two subsets, H_1 and H_2, of a space S constitute a *partition* of S if they are closed, non-null, disjoint, and have the union S. We write $S = H_1 \mid H_2$ for "$H_1 \mid H_2$ is a partition of S". H_1 and H_2 are also open sets, for each is the complement of the other; and since neither H_1 nor H_2 is null, neither can be identical with S. Conversely, if H is both open and closed in S, and is neither 0 nor S, $H \mid S - H$ is a partition.

The definition is carried over to sets of points in the usual way: $H_1 \mid H_2$ is a partition of E $(E = H_1 \mid H_2)$ if H_1 and H_2 are closed *in* E, non-null, disjoint, and have the union E.

A space, or set of points, is *connected* if it admits no partition. Thus the null-set and a set with only one point are connected, for they have no non-null subsets different from themselves; but $R^1 - (o)$ is not connected, for it is the union of the sets $[\xi > 0]$ and $[\xi < 0]$, which are closed in $R^1 - (o)$. An isolated space, S, with more than one point is not connected, for if a is any point, (a) and $S - (a)$ are closed sets.

A compact connected set with at least two points is called a *continuum*, a non-null open connected set in any space S is called a *domain*.

Connectedness is evidently a topologically invariant property. It is in fact invariant not only under topological mappings, but also under any continuous mapping. For suppose that S is connected but that $f(S) = H_1 \mid H_2$. By III, para. 9, $f^{-1}(H_1)$ and $f^{-1}(H_2)$ are non-null, disjoint, and closed, and their union is S, contradicting the assumption that S is connected.

Theorem 1·1. *R^1 is connected.*

Suppose that $R^1 = H_1 \mid H_2$, and that $\alpha \in H_1$ and $\beta \in H_2$, $(\alpha < \beta)$. The subsets of H_1 and H_2 in the compact set $< \alpha, \beta >$ are non-null, and they are closed and therefore compact. Hence they are at a positive distance δ apart, and contain points γ_1 and γ_2, respectively, such that $|\gamma_1 - \gamma_2| = \delta$. The point $\frac{1}{2}(\gamma_1 + \gamma_2)$ being within δ of both H_1 and H_2 can belong to neither, a contradiction of the definition of these sets.

Corollary. Since connectedness is a topological property it follows that the interval (α, β) is connected.

If $S = H_1 \,|\, H_2$, and E is any set in S, $E = EH_1 \,|\, EH_2$, provided that neither EH_1 nor EH_2 is null. For since H_1 and H_2 are closed, EH_1 and EH_2 are closed in E. This simple result is at the heart of many of the proofs that follow.

The following theorem plays a corresponding part in an alternative development of the theory, more "algebraical" than that here adopted. *A necessary and sufficient condition that the non-null sets H_1 and H_2 form a partition of $H_1 \cup H_2$ is that*

$$H_1 \bar{H}_2 \cup H_2 \bar{H}_1 = 0,$$

i.e. that neither of the sets H_1, H_2 contains a point or closure-point of the other. This follows easily from the fact that the condition that H_1 be closed in $H_1 \cup H_2$ is that

$$H_1 = \bar{H}_1(H_1 \cup H_2) = H_1 \cup \bar{H}_1 H_2.$$

Theorem 1·2. *If E is connected and $E \subseteq E_1 \subseteq \bar{E}$, then E_1 is connected.*

If $E_1 = H_1 \,|\, H_2$, either EH_1 or EH_2 is null, since otherwise $E = EH_1 \,|\, EH_2$. Suppose $EH_1 = 0$. Then $E \subseteq H_2$, $\bar{E} \subseteq \bar{H}_2$, and therefore, a fortiori, $E_1 \subseteq \bar{H}_2$. Hence

$$E_1 \subseteq \bar{H}_2(H_1 \cup H_2) = H_2,$$

for H_2 is closed in $H_1 \cup H_2$. This contradicts the assumption that $H_1 \neq 0$.

From this theorem and 1·1, Corollary it follows that $<\alpha, \beta>$, $<\alpha, \beta)$ and $(\alpha, \beta>$ are connected. Since $[\xi > \alpha]$ is the homeomorph of R^1 it is connected, and therefore its closure, $[\xi \geqslant \alpha]$, is connected. Similarly $[\xi < \alpha]$ and $[\xi \leqslant \alpha]$ are connected. The sets named in this paragraph, together with (α, β), $<\alpha, \beta>$, and R^1 itself, are called by the general name of *interval*, differentiated as open, closed and half-open.

The intervals are the only connected subsets of R^1 with more than one point. All connected subsets of R^1 are convex, for if E contains α and β, but not the point γ between them, the subsets of E in $[\xi < \gamma]$ and $[\xi > \gamma]$ form a partition of E. It follows that if α and β are the *l.u.b.* and *g.l.b.* of a connected set E, (the values $-\infty$ and ∞ being allowed) all points of (α, β) belong to E.

In particular, the only domains in R^1 are the sets (α, β), (α, ∞), $(-\infty, \alpha)$, and R^1 itself.

Example. If E_1 is the set of points

$$[\xi_2 = \sin 1/\xi_1, \quad 0 < \xi_1 \leqslant 1]$$

in R^2, and E_2 is any subset of the segment $[-1 \leqslant \xi_2 \leqslant 1]$ of the ξ_2-axis, the set $E_1 \cup E_2$ is connected. For E_1 is connected, since it is the homeomorph of $(0, 1>$, and $E_1 \subseteq E_1 \cup E_2 \subseteq \bar{E}_1$.

Exercise. Shew that if a continuous real function, f, in a connected space takes the values α and β it takes all intermediate values. By taking f to be $\rho(x, a)$ deduce that a connected set with more than one point is not enumerable. [Use II. 4·5.]

Theorem 1·3. *If a connected set of points in S meets both E and $S - E$ it meets $\mathscr{F}E$.*

Let the connected set be A. The whole space S is the union of the three disjoint sets $\mathscr{I}E$, $\mathscr{F}E$ and $\mathscr{I}(S - E)$. If $A\mathscr{F}E = 0$, A is the union of the sets $A\mathscr{I}E$ and $A\mathscr{I}(S - E)$, which are open in A. Neither of these sets is null, for if, for example, $A\mathscr{I}E$ is null, as well as $A\mathscr{F}E$,

$$A = A\mathscr{I}(S - E) \subseteq S - E,$$

i.e. A does not meet E.

We have therefore found a partition of A, contrary to the assumption that it is connected.

Two points x and y *are connected in* E if a connected subset of E contains them both.

Theorem 1·4. *If every two points of E are connected in E, E is a connected set.*

Suppose, if possible, that $E = H_1 \,|\, H_2$, and that $x_1 \in H_1$, $x_2 \in H_2$. Let E_0 be a connected subset of E containing x_1 and x_2. Then $E_0 H_1$ and $E_0 H_2$ are non-null, and therefore

$$E_0 = E_0 H_1 \,|\, E_0 H_2,$$

and is not connected.

Theorem 1·5. *If the connected sets E_a and E_b have a common point c, $E_a \cup E_b$ is connected.*

Suppose $E_a \cup E_b = H_1 \mid H_2$, and that $c \in H_1$. The set $E_a H_1$ is not null and therefore $E_a H_2 = 0$, since otherwise $E_a = E_a H_1 \mid E_a H_2$. Similarly $E_b H_2 = 0$, and therefore $H_2 = E_a H_2 \cup E_b H_2 = 0$: $H_1 \mid H_2$ is not a partition of $E_a \cup E_b$.

Corollary 1. *The union X of any system†️ of connected sets $\{E_z\}$ with a common point is connected*, for any two points of X belong to some set $E_a \cup E_b$.

Corollary 2. *If E is connected, and $\{E_z\}$ is a system of connected sets all meeting E, the set $X = E \cup \bigcup_z E_z$ is connected*, for any two points of X belong to some subset $E_a \cup E \cup E_b$, which is seen to be connected by two applications of 1·5.

From 1·4 it follows that all normed real vector spaces, and the space R^ω, and all linear or convex subsets of these spaces, are connected, for every pair of their points is connected by a segment, which, as the homeomorph of $< 0, 1 >$ in R^1, is connected. The space Z^p is connected, since its dense subset R^p is connected; and hence S^p is connected.

The following criterion gives expression to the intuitive notion of connection as the property of being joined by short steps. A *chain of sets* is a finite series of sets $E_1, E_2, ..., E_k$ such that E_i meets E_{i+1} $(i = 1, 2, k-1)$.

Theorem 1·6. *A necessary and sufficient condition for S to be connected is that, given any open covering, or finite closed covering, $\{E_z\}$, of S, any two members of $\{E_z\}$ are the first and last members of a chain of sets E_z.*

Necessary. If E_a is a set E_z, let H_1 be the union of those sets E_z that are joined to E_a by a chain of sets E_z, and H_2 the union of all the rest. Then H_1 and H_2 are both open or both closed; they clearly cannot meet; and $H_1 \supseteq E_a \neq 0$ (by the definition of a covering).‡ Hence $H_2 = 0$, i.e. every E_z is joined to E_a by a chain.

Sufficient. The defining property of connection is the special case of either form of the given condition where the covering has two members.

---

† "Systems of sets" is used as an alternative for "set of sets", for linguistic convenience.   ‡ The members of a covering are by definition non-null.

Theorem 1·7. *If a locally separable†️ space is connected, it is separable.*

For each x let $\delta(x)$ be the *l.u.b.* of numbers δ such that $U(x, \delta)$ is separable (we may suppose the space to have diameter 1). Let a dense enumerable set $A(x)$ be chosen in each neighbourhood $U(x, \frac{1}{2}\delta(x))$. Let $E_1 = A(a)$ for some point a, and, supposing E_{n-1} already defined, let

$$E_n = \bigcup_{x \in E_{n-1}} A(x), \quad A = \bigcup_{n=1}^{\infty} E_n.$$

The sets E_n and A are clearly enumerable. Let $x \in \bar{A}$. For some n there exists a point a_n of E_n within $\frac{1}{5}\delta(x)$ of x. Since

$$U(a_n, \tfrac{4}{5}\delta(x)) \subseteq U(x, \delta(x))$$

we have $\delta(a_n) \geqslant \frac{4}{5}\delta(x)$, and

$$U(x, \tfrac{1}{5}\delta(x)) \subseteq \overline{U(a_n, \tfrac{2}{5}\delta(x))} \subseteq \overline{U(a_n, \tfrac{1}{2}\delta(a_n))}$$
$$= \bar{A}(a_n) \subseteq \bar{E}_{n+1} \subseteq \bar{A}.$$

Thus the separable set $\bar{A}$ is open as well as closed, and must therefore be the whole space.

2. One of the central problems of topology is to discover necessary and sufficient conditions for homeomorphism between two spaces. There are two sides to this problem—the condition that the spaces should be homeomorphic, and the condition that they should not. Little progress has been made on the positive side beyond the classification of surfaces (2-dimensional manifolds), but any topologically invariant property is a means of shewing that certain pairs of spaces are *not* homeomorphic, for if S has the property and T has not, S and T are not homeomorphic. Compactness and local compactness have already been used in Chapter III to distinguish certain pairs of spaces. In *connection* we have the material for more delicate tests, by considering the effect on the connection of a space of removing a set of some prescribed character. This is the method underlying the proof, in Chapter V, of the theorem that, if $p \neq q$, R^p and R^q are not homeomorphic. Some simpler applications are given in the rest of this paragraph.

† I.e. a space in which every point has a separable neighbourhood.

Theorem 2·1. *If $p > 1$ the connection of R^p is not destroyed by removing an enumerable set of points, E.*

Let x and y be any points of $R^p - E$, and l a straight line meeting the segment xy but not passing through x or y. If z_1 and z_2 are distinct points of l, the only common points of $xz_1 \cup z_1 y$ and $xz_2 \cup z_2 y$ are x and y. Therefore if, for every z of l, E meets $xz \cup zy$, E must contain a subset which is in $(1, 1)$-correspondence with the set of points l, and therefore not enumerable. This is impossible, since E is enumerable. Therefore, for at least one z, $xz \cup zy$ does not meet E, i.e. x and y are connected in $R^p - E$.

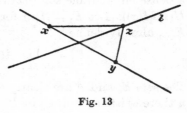

Fig. 13

From this it follows that *if $p \neq 1$, R^p and R^1 are not homeomorphic.* For the connection of R^1 is destroyed by the removal of a single point.

A set of points which is the homeomorph of a segment is called a *simple arc*, and the homeomorph of a circle is called a *simple closed curve*, or *Jordan curve*. A simple arc is not the homeomorph of a simple closed curve because the connection of a segment (and therefore of any simple arc) is destroyed by removing any point, with two exceptions (the end-points); but the connection of a circle (or other simple closed curve) cannot be destroyed by the removal of one point.

A simple arc or simple closed curve in R^p is nowhere dense, if $p > 1$. A set of points of a circle, or of a segment, is either not connected, or its connection is destroyed by the removal of any three points. The same is therefore true of a set of points of a simple closed curve or simple arc. But if a simple arc or simple closed curve, E, in R^p is *not* nowhere dense it contains a circular neighbourhood, U, and if $p > 1$ the connection of U is not destroyed by the removal of any finite subset. Therefore E is nowhere dense.

Exercises. 1. It was shewn above that the only non-null connected subsets of R^1 are (for arbitrary α and β, subject to $\alpha < \beta$):

(a) R^1, (α, β), $(-\infty, \alpha)$, (α, ∞);

(b) $(\alpha,\beta>$, $<\alpha,\beta)$, $(-\infty,\alpha>$, $<\alpha,\infty)$;
(c) $<\alpha,\beta>$.

Prove that the sets in the three rows are of three topologically different types, and that all of them are different topologically from a circle. [The connection of (a), (b), and (c) is destroyed by the removal of one point, which may in case (a) be any point, in case (b) any point with one exception and in case (c) any with two exceptions.]

2. If $p>1$, no subset of R^1 or of S^1 is the homeomorph of R^p.

3. If $p>1$, S^1 and S^p are not homeomorphic.

4. If, in the example on p. 73, E_2 contains at least two points, $E_1 \cup E_2$ is not the homeomorph of $<0,1>$. (It will be shewn in §4, Example 4, p. 85, that this is still true if E_2 is a single point.)

§ 2. COMPONENTS

3. If E is a set of points in any space, the relation between any two points of being *connected in* E is an equivalence relation. For it is obviously symmetrical and reflexive, and if the pairs of points x and y, y and z are connected in E by the subsets A and B respectively, $A \cup B$ is a connected subset of E containing x and z: the relation is transitive. The set of points E therefore falls into disjoint subsets such that two points are connected in E if, and only if, they belong to the same subset. These subsets are called the *components* of E. (They are by definition not null.)

Every connected subset, A, of E is contained in a single component of E, for any two of its points are connected in E by A itself. It follows that *components are connected sets*, for the connected subset of E joining any two points of a component C lies entirely in C. Thus a necessary and sufficient condition that C be a component of E is that it be a maximal connected subset, i.e. one not contained in any other connected subset.

If C is a component of a space S, $\bar{C}$ is a connected subset of the space meeting C, and is therefore contained in C. Thus C is closed. It follows that *the components of any set E are closed in E*. Hence *the components of a closed set are closed sets* (in S).

Examples. The components of $R^1-(o)$ are the sets $[\xi>0]$ and $[\xi<0]$, the components of the sets $(1, \frac{1}{2}, \frac{1}{3}, \ldots)$ and Q^1 (the rational points) in R^1 are the individual points of these sets.

Theorem 3·1. *If a domain D is a component of the open set G, then $\mathscr{F}D \subseteq \mathscr{F}G$.*

D is closed in G, i.e. $D = \bar{D}G$. Hence

$$\mathscr{F}D = \bar{D} - D = \bar{D} - \bar{D}G$$
$$= \bar{D}(S - G) \subseteq \bar{G}(S - G) = \mathscr{F}G.$$

Theorem 3·2. *If D is a domain it is a component of $S - \mathscr{F}D$.*

We have to shew that the component of $S - \mathscr{F}D$ containing D coincides with D. If it does not it is a connected set meeting both D and $\mathscr{C}D$, and it therefore meets $\mathscr{F}D$. This is impossible, since it is a subset of $S - \mathscr{F}D$.

Theorem 3·3. *If A is a connected subset of a connected space S, and C is any component of $S - A$, then $S - C$ is connected.*

(The example in Fig. 14 illustrates the theorem and its proof.
S is the whole plane, A the white
domain and C the *inner* shaded
region.)

Let H be a subset of $S - C$
that is both open and closed in
$S - C$. Then $C \cup H$ *is connected*.
For if $H_1 \mid H_2$ is a partition of
$C \cup H$, C is contained in one
part, say H_1, and therefore
$H_2 \subseteq H$. Since H_2 is open and
closed in $C \cup H$ it is open and
closed in the smaller set H.

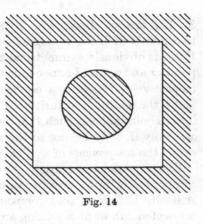

Fig. 14

Therefore (II. 16·3 and 16·4), since H is open and closed in $S - C$, H_2 is open and closed in $S - C$. Hence, finally, by II. 16·5, H_2 is open and closed in

$$(S - C) \cup (C \cup H) = S,$$

contrary to the assumption that S is connected.

Now suppose, if possible, that $S - C = H_3 \mid H_4$. If AH_3 were null, $C \cup H_3$, which, as we have just shewn, is connected, would lie in $S - A$. Since $C \cup H_3$ contains C as a *proper* subset (for $CH_3 = 0$ and $H_3 \neq 0$), this contradicts the definition of C as a component of $S - A$. Hence AH_3, and similarly AH_4, is not null, i.e. $A = AH_3 \mid AH_4$, contrary to the assumption that A is connected.

Theorem 3·4. *If E is any connected subset of a connected space, S, and H is open and closed in $S - E$, then $E \cup H$ is connected.* (Proved in the course of 3·3.)

A theorem resembling 3·3, but with $\mathscr{F}C$ in place of $S - C$, will be shewn in Part II to hold for closed sets, A, in R^p and Z^p. That the theorem, so altered, does not hold, e.g. on a ring-surface, is shewn by Fig. 15 (A is the shaded region).

4. The components of an open set in a metrisable space of general character need not be open; in particular the components of the space itself need not be open sets. The symmetry between "closed" and "open" breaks down at this point. For example, the point (o) is a component of the set Q^1, (the rational points of R^1), but $Q^1 - (o)$ is not closed in Q^1. Thus if C is a component of a set E, C and $(E - C)$ do not necessarily form a partition. It is not even

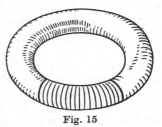

Fig. 15

true in general that if x and y are not connected in E there is a partition of E such that x and y belong to different parts.

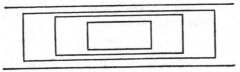

Fig. 16

Example (Hausdorff): Let E consist of the two lines $\xi_2 = \pm 1$, and all the rectangles† A_n with centre o and sides, of length $2n$ and $2 - 1/n$, parallel to the ξ_1- and ξ_2-axes respectively ($n = 1$, $2, \ldots$). *Each of the two straight lines $\xi_2 = \pm 1$ is a component.* For if the component, C, of E containing the line $\xi_2 = 1$ contained a point of A_n,

$$A_n \mid C - A_n$$

would be a partition of C. Therefore C is contained in the two

† I.e. the perimeters of these rectangles.

straight lines $\xi_2 = \pm 1$, and since it is connected must be $\xi_2 = 1$ itself. The points x and y in which the ξ_2-axis cuts the two lines $\xi_2 = \pm 1$ are therefore not connected in E, but *in any partition they belong to the same part.* For at least one part contains an infinity of rectangles, and therefore, since it is closed, both the lines $\xi_2 = \pm 1$.

Another unexpected possibility is illustrated in the example below. This and other examples that may be constructed shew that the components of a space may lack many of the properties that intuition suggests they should have, and it is better, in dealing with spaces of general character, to work as far as possible with partitions rather than with components. There are, however, two kinds of spaces in which these difficulties largely disappear, the *compact* spaces and the *locally connected* spaces. They are considered in §§ 3 and 4 of this chapter.

Example. If a is a point of a connected set E, and C a component of $E - (a)$, does a necessarily belong to $\overline{C}$? The consideration of any "natural" example suggests that it must, but it will now be shewn that this is not correct. Let E_1 be the union of the segments in R^2 joining o to the points

$$(1, 1), \ (1, \tfrac{1}{2}), \ (1, \tfrac{1}{3}), \ \ldots,$$

and E_2 the segment $< \tfrac{1}{2}, 1 >$ of the ξ_1-axis. Then E_1 is evidently connected, and since

Fig. 17

$$E_1 \subseteq E_1 \cup E_2 \subseteq \overline{E}_1,$$

$E_1 \cup E_2$ is connected. But in $(E_1 \cup E_2) - (o)$ the remains of all the radiating segments are separate components, and the segment E_2 is also a component.

Exercises. 1. Shew that a set which has only a finite number of different partitions has only a finite number of components.

2. Shew by an example that in 3·4 "H is open and closed in $S - E$" cannot be replaced by "H is a component of $S - E$".

§ 3. CONNECTION IN COMPACT SPACES

5. In this section a metric ρ is to be chosen in the topological space, and the truth of the theorems is asserted for every possible choice of ρ.

In compact spaces the criterion of "connection by short steps" (1·6) takes a simpler form. An ϵ-*chain* is a finite succession of points, $a_1, a_2, ..., a_q$ such that $\rho(a_i, a_{i+1}) \leqslant \epsilon$, for $i = 1, 2, ..., q - 1$. A space is ϵ-*connected* if every pair of points in it can be joined by an ϵ-chain of points in the set. It follows from Theorem 1·6, on taking the open covering to be the set of all $\frac{1}{2}\epsilon$-neighbourhoods, that a connected space is ϵ-connected for every positive ϵ. The converse, which is not true in general (as shewn by the example of $R^1 - (o)$), holds in compact spaces.

Theorem 5·1. *A necessary and sufficient condition for a compact space S to be connected is that it be ϵ-connected for every positive ϵ.*

Only the sufficiency remains to be proved. Suppose that $S = H_1 \mid H_2$. Then $\rho(H_1, H_2) = \epsilon > 0$ since H_1 and H_2 are compact. Clearly no $\frac{1}{2}\epsilon$-chain can join a point of H_1 to a point of H_2, and the space is therefore not $\frac{1}{2}\epsilon$-connected.

The theory of connection in compact sets is greatly simplified by this result.

Theorem 5·2. (Lemma.) *The common part, F, of a decreasing sequence of compact, ϵ-connected sets is 2ϵ-connected.*†

Let $F_1, F_2, ...$ be the sequence of sets, F their common part, and a, b any two points of F. Let n be such that $F_n \subseteq U(F, \frac{1}{2}\epsilon)$, and $a = a_0, a_1, ..., a_k = b$ an ϵ-chain in F_n. Let b_i be, for $1 \leqslant i \leqslant k - 1$, a point of F in $U(a_i, \frac{1}{2}\epsilon)$. Then $a, b_1, b_2, ..., b_{k-1}, b$ is a 2ϵ-chain in F.

The lemma has two important applications.

Theorem 5·3. *The common part, F, of a decreasing sequence of continua is a continuum or a point.*

That F is not null was shewn in II. 15·4. Given any positive ϵ the sets F_n are all ϵ-connected. Therefore F is 2ϵ-connected for every positive ϵ, and therefore it is connected.

---

† It can be shewn that F is in fact ϵ-connected.[11]

Theorem 5·4. *If, for every positive ϵ, the points a and b of the compact set F can be joined by an ϵ-chain in F they are connected in F.*

(This is a deeper theorem than 5·1: compare the example on p. 79.)

Let E_n be the set of points of F that can be joined to a by a $1/n$-chain in F. By hypothesis b belongs to E_n for every n, and therefore also to $\cap E_n$. The set $F - E_n$ is open, for if $x \in F - E_n$, evidently all points of $U(x, 1/n)$ are in $F - E_n$. Therefore the sets E_n are closed, and hence compact; they form a decreasing sequence; and E_n is $1/n$-connected, and therefore $1/n_0$-connected if $n > n_0$, where n_0 is any fixed integer. Hence, by 5·2,

$$\overset{\infty}{\underset{1}{\cap}} E_n = \overset{\infty}{\underset{n_0}{\cap}} E_n$$

is $2/n_0$-connected for every n_0, and therefore connected.

Thus a and b are connected in F by the set $\cap E_n$.

Theorem 5·5. *If a and b are points of the compact set F which are not connected in F, there is a partition $H_1 \mid H_2$ of F such that $a \in H_1$ and $b \in H_2$.*

By the preceding theorem there is a positive number ϵ such that a and b cannot be joined in F by an ϵ-chain. Let H_1 be the set of points of F that can be joined to a by an ϵ-chain and let $H_2 = F - H_1$. Since

$$\rho(H_1, H_2) \geqslant \epsilon > 0,$$

$H_1 \mid H_2$ is a partition, and $a \in H_1$, $b \in H_2$.

This theorem has the following generalisation:

Theorem 5·6. *If X and Y are non-null closed subsets of a compact set F, such that no component of F meets both of them, there is a partition $H_1 \mid H_2$ of F such that $X \subseteq H_1$, $Y \subseteq H_2$.*

We first prove that *there is a positive number η such that no point of X is joined to a point of Y by an η-chain in F.* If the assertion is false let x_n and y_n be, for each n, points of X and Y joined by a $1/n$-chain in F. A subsequence x_{m_r} of the x_n converges to a point x of X, and a subsequence y_{n_r} of the y_{m_r} converges to a point y of Y. Given any positive ϵ, a number r can be found such that

$$\frac{1}{n_r} < \epsilon, \quad \rho(x, x_{n_r}) < \epsilon, \quad \rho(y, y_{n_r}) < \epsilon;$$

and $\qquad\qquad x, \ \dfrac{1}{n_r}$-chain from x_{n_r} to y_{n_r}, $\ y$

is an ϵ-chain from x to y. Therefore x and y are connected in F, contrary to the assumption that no component of F meets both X and Y.

The proof now follows that of 5·5, H_1 being the set of points of F that can be joined to some point of X by an η-chain.

Corollary 1. *If C is a component of a compact set F, and ϵ is positive, there is an open and closed subset H of F such that $C \subseteq H \subseteq U(C, \epsilon)$.* If $F \subseteq U(C, \epsilon)$ the required set H is F itself. If not let $X = C$, $Y = F - U(C, \epsilon)$, $H_1 = H$ in Theorem 5·6.

Corollary 2. *In a connected, locally compact space, S, containing more than one point, every point is contained in an arbitrarily small continuum.*† Let $U(a, \epsilon)$ be a neighbourhood whose closure is compact. We may evidently assume that $U \neq S$. If a is a one-point component of $\overline{U}$, there is, by Corollary 1, a set, H, open and closed in $\overline{U}$, such that $a \in H \subseteq U(a, \epsilon)$. Thus H is open in U, and therefore in S; it is closed in $\overline{U}$, and therefore in S; and $H \neq S$. Since S is connected there is no such set H, and the a-component of $\overline{U}$ is the required continuum.

Theorem 5·7. *If x_n and y_n are, for each n, points of a compact set F such that* (1) *x_n and y_n are connected in F, and* (2) *$x_n \to x$ and $y_n \to y$, then x and y are connected in F.*

If x and y are not connected in F let ϵ be a positive number such that they are not joined by any ϵ-chain in F. Let n be such that $\rho(x, x_n) < \epsilon$ and $\rho(y, y_n) < \epsilon$. The points x_n and y_n can be joined by an ϵ-chain in F, and this chain, preceded by x and followed by y, is an ϵ-chain joining x and y in F, contrary to the definition of ϵ.

Example. 1. *If F_1 and F_2 are compact sets, the union, X, of the components of F_1 that meet F_2 is closed in F_1.* Let (x_n) be a sequence of points of X converging to x of F_1. Let the component of F_1 containing

† A continuum contains more than one point.

x_n meet F_2 in y_n, and let $y_{n_r} \to y$ in F_2. Then, by 5·7, x and y are connected in F_1. Thus x belongs to a component of F_1 that meets F_2, and is therefore in X.

Exercise. If F_0 is a continuum, F_1 a closed subset of F_0, and C a component of $F_0 - F_1$, then $\bar{C}$ meets F_1. [If $\rho(F_1, C) = \epsilon > 0$, 5·6 Corollary 1 gives a set H open and closed in $F_0 - U(F_1, \frac{1}{2}\epsilon)$ and satisfying $C \subseteq H \subseteq U(C, \frac{1}{2}\epsilon)$. Then $H \mid F_0 - H$ is a partition of F_0. (This shews that in compact sets the answer to the question in the Example, p. 80, is "Yes".)]

The points x and y of a space S are *strongly connected* in S if they belong to a compact connected subset of S; and S itself is strongly connected if every two of its points are. Being strongly connected is an equivalence relation, and the classes into which it divides the points of S are sometimes called *constituents*. They need be neither open nor closed.

Example. 2. Let E_1 be formed from the set E in Fig. 16 by adjoining the ξ_1-axis. Then E_1 is connected, but not strongly connected, for the points $(0, \pm 1)$ are not joined by any continuum. The "constituents" are (1) the union of all the rectangles and $\xi_2 = 0$ (open but not closed), (2) and (3) the lines $\xi_2 = \pm 1$ (closed but not open).

§ 4. LOCAL CONNECTION

6. A space S is *locally connected at a* if, given any positive ϵ, there exists a positive δ such that any two points of $U(a, \delta)$ are joined by a connected set in $U(a, \epsilon)$. The space is *locally connected* if it is locally connected at each of its points.[12]

Theorem 6·1. *All real normed vector spaces and the space R^ω, and all linear or convex subsets of these spaces, are locally connected.*

For in any such space or set, two points x and y of a neighbourhood $U(a, \epsilon)$ are joined in the neighbourhood itself by the pair of segments $xa \cup ay$.

Theorem 6·2. *Any open set in a locally connected space is itself locally connected.*

If $a \in G \subseteq S$, and $GU(a, \epsilon)$ is a neighbourhood relative to G, there exists $\epsilon_1 \leqslant \epsilon$ such that $U(a, \epsilon_1) \subseteq G$. If δ_1 corresponds to ϵ_1 as in the definition of local connection, any two points of $U(a, \delta_1)$ are connected in $U(a, \epsilon_1)$, and a fortiori in $GU(a, \epsilon)$.

The following examples shew how a set can fail to be locally connected.

Examples. 1. The set of points $\{0, 1, \frac{1}{2}, \frac{1}{3}, \ldots\}$ in R^1 is not locally connected, since in any neighbourhood of o there are points of the set not connected with o in the set.

2. Let the origin in R^2 be joined by a segment to each of the points $(1, 1/n)$, for positive integral n, and to $(1, 0)$. This set, E_0, is not locally connected at points $(\alpha, 0)$ $(0 < \alpha \leqslant 1)$. For any connected subset of E_0 containing the points $(\alpha, 0)$ and $(\alpha, \alpha/n)$ must evidently contain the whole of the segment from o to $(\alpha, \alpha/n)$, and has therefore diameter at least α; but by making n large enough $(\alpha, \alpha/n)$ can be brought arbitrarily near $(\alpha, 0)$.

3. If in the previous example any subset of the points on the ξ_1-axis is omitted the new set is still not locally connected provided at least one point $(\alpha, 0)$ with $\alpha > 0$ remains.

4. The set of points $E_1 \cup E_2$, where E_1 is the set

$$[\xi_2 = \sin 1/\xi_1, \quad 0 < \xi_1 \leqslant 1]$$

and E_2 any non-null subset of the segment $< -1, 1 >$ of the ξ_2-axis, is not locally connected at points of E_2.

5. The set given in plane polar coordinates by

$$[\theta = (1-r)^{-1}, \quad 0 \leqslant r < 1] \cup [r = 1, \quad \text{all } \theta],$$

i.e. a circle and a curve approaching it spirally from within, is not locally connected at points of the circle.

Fig. 18.   Example 2

Theorem 6·3. *A necessary and sufficient condition for S to be locally connected at a is that in every neighbourhood U_a there is a connected set X having a as an interior point.*

Necessary. Let $U(a, \epsilon)$ be a neighbourhood of a, δ a number corresponding to ϵ as in the definition of local connection, and for each x of $U(a, \delta)$ let E_x be a connected set in $U(a, \epsilon)$ containing a and x. Then if

$$X = \bigcup_{x \in U(a, \delta)} E_x,$$

X is connected, it is contained in $U(a, \epsilon)$, and since $U(a, \delta) \subseteq X$, $a \in \mathscr{I} X$.

Sufficient. Given any $U(a, \epsilon)$ let X be a connected set in $U(a, \epsilon)$ such that $a \in \mathscr{I}X$. There is then a neighbourhood $U(a, \delta)$ contained in X, any two points of which are connected in $U(a, \epsilon)$ by X itself.

From this result it is clear that the property of being locally connected is a topological one, independent of the particular metric used in defining it.

Theorem 6·4. *A necessary and sufficient condition for a space to be locally connected is that the components of all open sets are open.*

Necessary. If C is a component of an open set G, and $a \in C$, let $U_a \subseteq G$. By 6·3 a connected subset X of U_a has a as an interior point. Since X is a connected subset of G and meets C, $X \subseteq C$. Hence $a \in \mathscr{I}C$, and C is therefore open.

Sufficient. If a is any point and U_a any of its neighbourhoods the component of U_a containing a is the connected set X of 6·3.

Corollary. *A necessary and sufficient condition for a space to be locally connected is that it have a base of connected open sets.* Necessary: the components of all open sets from such a base. Sufficient: by 6·3.

Example. 6. Consider the part of the set E_0 (Example 2 above, Fig. 18), inside the circle with centre $(\frac{1}{2}, 0)$ and radius $\frac{1}{4}$. It is open in E_0, since it is the intersection of E_0 with an open set in R^2. One of its components is the interval $(\frac{1}{4}, \frac{3}{4})$ of the ξ_1-axis. *This component is not open in E_0,* for the point $(\frac{1}{2}, 0)$ in it is a limit-point of points of E_0 not on the ξ_1-axis.

An important special case of 6·4 is that *the components of a locally connected space, S, are open sets.* Therefore every component, C, of S is both open and closed, i.e. $C \mid (S - C)$ *is a partition of the space S.* It was the failure of this property in general spaces that led to the difficulties mentioned on p. 80.

Theorem 6·5. *Let E be any set in a locally connected space S. If C is a component of E, $\mathscr{I}C = C \cap \mathscr{I}E$; and if E is closed, $\mathscr{F}C = C \cap \mathscr{F}E$.*

Any component of $\mathscr{I}E$ that meets C is contained in C, and therefore $C\mathscr{I}E$ is the union of components of the open set $\mathscr{I}E$, i.e. it is an open set contained in C. Therefore $C\mathscr{I}E \subseteq \mathscr{I}C$; and since $C \subseteq E$, $\mathscr{I}C = C\mathscr{I}C \subseteq C\mathscr{I}E$.

If E is closed, C is closed; and
$$\mathscr{F}C = C - \mathscr{I}C = C(E - \mathscr{I}E) = C\mathscr{F}E.$$

If X is any locally connected set (e.g. any open set) in the subspace $[\xi_1 = 0]$ of R^p, and $\phi(\xi_2, ..., \xi_p)$ is a continuous real function in X, the graph $[\xi_1 = \phi]$ of ϕ is locally connected, for it is homeomorphic with X (Example 8, p. 63). It follows that if a function $\psi(\xi_1, \xi_2, ..., \xi_p)$ is such that in some neighbourhood of each point of $[\psi = 0]$, the equation $\psi = 0$ determines at least one coordinate ξ_k as a single-valued continuous function of the others, then the set $[\psi = 0]$ is locally connected. This suffices to prove the local connection of many familiar sets in R^p, including S^{p-1}.

The mere continuity of ψ is not sufficient to ensure that $[\psi = 0]$ is locally connected: indeed if F is any closed set in any space S, $\rho(x, F)$ is a continuous function in S, and $[\rho = 0]$ is F.

7. An admissible metric ρ in a topological space is said to be *locally connected* if, for some positive ϵ_0, all neighbourhoods $U_\rho(x, \epsilon)$ of radius less than ϵ_0 are connected.

Theorem 7·1. *A necessary and sufficient condition for a space to be locally connected is that it admit a locally connected metric.*

Necessary. Let ρ be any admissible metric, and let $\eta(x, y)$ be defined for all x and y as follows. If x and y are connected in S by a set of diameter less than 1, let $\eta(x, y)$ be the greatest lower bound of the diameters of all such sets; in all other cases let $\eta(x, y) = 1$. Thus $\eta(x, y) \leqslant 1$ for all x and y.

The function η satisfies the three conditions m_1, m_2, m_3 for a metric (p. 17). That it satisfies m_1 and m_2 is obvious. If x, y and z are any three points, then if either $\eta(x, y)$ or $\eta(y, z)$ is 1,
$$\eta(x, y) + \eta(y, z) \geqslant 1 \geqslant \eta(x, z).$$
If $\eta(x, y)$, $(= \alpha)$, and $\eta(y, z)$, $(= \beta)$, are both less than 1, there exist connected sets, A and B, of diameter less than $\alpha + \epsilon$ and $\beta + \epsilon$, joining x and y, y and z respectively. $A \cup B$ is a connected set of diameter less than $\alpha + \beta + 2\epsilon$, joining x and z. Since ϵ is arbitrary it follows that
$$\eta(x, z) \leqslant \alpha + \beta = \eta(x, y) + \eta(y, z),$$
i.e. m_3 is satisfied.

88 TOPOLOGY OF SETS OF POINTS IV. 7

η and ρ are equivalent metrics. It is sufficient to shew that η and ρ^* (p. 58) are equivalent. Since $\rho^*(x,y) \leqslant \eta(x,y)$ for all x and y, the first half of the conditions of III. 1·2 is satisfied. That the second half ($\eta \to 0$ as $\rho^* \to 0$) be satisfied is precisely the condition for local connection.

The neighbourhood $U_\eta(a,\epsilon)$ is connected if $\epsilon < 1$. If $x \in U_\eta(a,\epsilon)$, x is connected to a by a set E of ρ-diameter less than ϵ. Every point of E is also connected to a by a set of ρ-diameter less than ϵ, namely E itself, and therefore $E \subseteq U_\eta(a,\epsilon)$. Thus x is connected to a in $U_\eta(a,\epsilon)$, and therefore $U_\eta(a,\epsilon)$ is connected.

Sufficient. If the space admits a locally connected metric, we can, for any positive ϵ, take δ to be min $(\epsilon, \frac{1}{2}\epsilon_0)$ in the definition of local connection.

Exercise. If E is a connected subset of a locally connected space and $\epsilon > 0$, there exists a connected open set G satisfying $E \subseteq G \subseteq U_\rho(E, \epsilon)$ (ρ being any metric).

8. Spaces that are compact and locally connected form an important class.

Theorem 8·1. *A necessary and sufficient condition for a space S to be compact and locally connected is that (relative to any admissible metric, ρ) it have a finite ϵ-covering by compact connected sets, for every positive ϵ.*

Necessary. If the space is locally connected, the components of all $\frac{1}{2}\epsilon$-neighbourhoods form an open covering from which, if S is compact, a finite covering $\{G_i\}$ can be selected. $\{\bar{G}_i\}$ is the required ϵ-covering.

Sufficient. The condition evidently implies compactness. Let ϵ be any positive number, and $\{K_i\}$ a finite ϵ-covering by connected compact sets. Let a be any point and δ_a the distance of a from the union of the K_i's that do not contain a (or if they all contain a, let $\delta_a = 1$). Then $\delta_a > 0$; and if $\rho(x,a) < \delta_a$, x belongs to a K_i containing a. Hence S is locally connected at a.

Corollary. *If X and Y are compact and locally connected, so also is $X \cup Y$.* (Counter-example without compactness: 4, p. 85.)

Theorem 8·2. *The image T of a compact locally connected space S under any continuous mapping f is compact and locally connected.*

Given any positive ϵ, let δ be positive and such that if $\rho(x,y) < \delta$ in S, $\sigma(fx,fy) < \epsilon$ in T. Then if $\{F_i\}$ is a finite δ-covering of S by connected compact sets, $\{f(F_i)\}$ is a finite ϵ-covering of T by connected compact sets. Hence T is compact and locally connected.

Exercise. Shew by an example that 8·2 does not remain true if "compact" is omitted at both its occurrences.

9. The famous theorem of Hahn and Mazurkiewicz will now be proved, that the segment $< 0, 1 >$ can be mapped continuously on to (the whole of) any locally connected continuum F.† The surprising nature of this result can be seen from the fact that it includes as a special case the theorem (Peano, 1890) that a segment can be mapped on to a square; or, in more concrete terms, a simple arc can be so disposed in ordinary 3-space that its shadow under parallel rays falling on a plane is a square. The following brief sketch of Hilbert's proof of this case may help to elucidate the proof of the general theorem.

If the segment is interpreted as a time-interval, say of one second, during which a point is to move over the square, we agree first, that the whole interval is to be spent somewhere in the square; next, that the successive quarter-seconds are to be spent in the quarter-squares 1, 2, 3, 4, in that order; next, that the sixteenth-seconds are to be spent in 11, 12, 13, 14, 21, 22, 23, 24, 31, 32, 33, 34, 41, 42, 43, 44, successively; and so on. Every instant τ is the limit of some decreasing sequence of time-intervals, of length 4^{-m}

Fig. 19

$(m = 1, 2, \ldots)$, and the position of $f(\tau)$ is the limit of the corresponding descending sequence of squares, of side 2^{-m}. If the

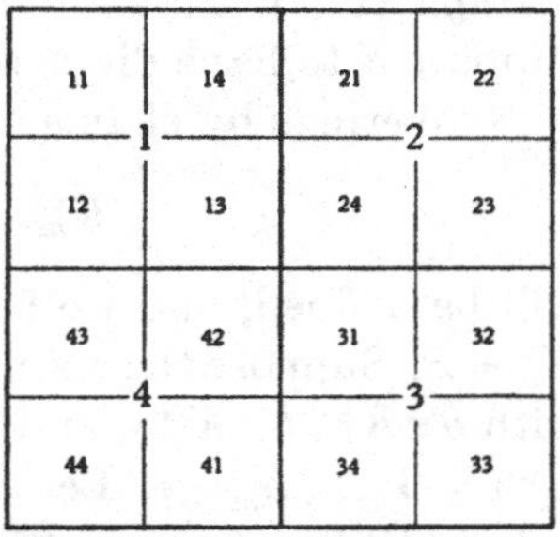

† The name "Peano continuum" or "Peano curve" is sometimes used for locally connected continuum.

N T

numbering at stage $(m+1)$ is so chosen that the last quarter of
any square of order m is adjacent to the first quarter of the
next square of order m (as is always possible), the region
allotted to any sufficiently small time-interval is arbitrarily
small, and hence f is continuous.

The details of this outline will not be filled in, since the proof
of 9·2 is a direct generalisation, the sets K_m^i replacing the squares
of order m.

9·1. (Lemma.) *Any finite closed covering $\{F_i\}$ of a connected set
can be arranged as a chain of sets, beginning and ending with assigned
sets F_a, F_b, provided that repetitions of the same set are allowed.*

Let the sets $\{F_i\}$ be arranged in any order $F_a = F_1, F_2, ..., F_k = F_b$.
By 1·6, F_r and F_{r+1} are, for each r, joined by a chain of the sets
$\{F_i\}$. On inserting these chains between the original consecutive
pairs, the required chain is obtained.

9·2. *A necessary and sufficient condition for a space to be the
continuous image of $<0,1>$ under some mapping f is that it be
connected, compact, and locally connected.*

Necessary. Follows immediately from 8·2, since the image of
a connected set is connected.

Sufficient. Let a and b be any points of the space S. We may
suppose S to have diameter $\leqslant 1$. A series of integers N_m, and of
2^{-m}-coverings by chains of connected compact sets

$$K_m^i \quad (i = 1, 2, ..., N_m)$$

will be defined, and we begin the induction by putting $N_0 = 1$,
$K_0^1 = S$. Suppose the definition completed up to and including m,
with $a \in K_m^1$, $b \in K_m^{N_m}$, and let a point a_i be chosen in $K_m^i \cap K_m^{i+1}$,
for $i = 1, ..., N_m - 1$. Let $a_0 = a$, $a_{N_m} = b$.

Let $\{F_r\}$ be a finite $2^{-(m+1)}$-covering of S by connected compact
sets. The sets $F_r \cap K_m^i$ that are not null form a finite closed
covering of K_m^i, and hence can be arranged (with repetitions) as
a chain of sets, beginning with a set containing a_{i-1} and ending
with one containing a_i. The corresponding sets F_r a fortiori form
a chain covering K_m^i. If n_m is the greatest number of sets

occurring in any one of these chains we may suppose they all contain exactly n_m sets, by suitable repetitions of the last member. The chains of sets associated in this way with

$$K_m^1,\ K_m^2,\ ...,\ K_m^{N_m},$$

placed consecutively in this order, form a single chain of connected closed sets which we suppose numbered through as

$$K_{m+1}^i \quad (i = 1, 2, ..., n_m N_m).$$

This completes the recursive definition of K_m^i and $N_{m+1} = n_m N_m$.

If $0 < j \leqslant N_m$, and

(1) $$n_m(j-1) < i \leqslant n_m j$$

the sets K_m^j and K_{m+1}^i meet, and therefore, since $\Delta(K_{m+1}^i) \leqslant 2^{-(m+1)}$, $K_{m+1}^i \subseteq U(K_m^j, 2^{-(m+1)})$. Hence if A_m^i denotes $U(K_m^i, 2^{-m})$ we have $A_{m+1}^i \subseteq A_m^j$ whenever i and j satisfy (1).

If $0 \leqslant \xi < 1$ there is a unique integer j $(0 < j \leqslant N_m)$ such that

$$(j-1)/N_m \leqslant \xi < j/N_m.$$

We denote by $A_m(\xi)$ the set A_m^j for this value of j; and define $A_m(1)$ to be $A_m^{N_m}$. If, for the same ξ, $A_{m+1}(\xi) = A_{m+1}^i$, i.e. if also

$$(i-1)/N_{m+1} \leqslant \xi < i/N_{m+1},$$

we must have

$$(j-1)/N_m \leqslant (i-1)/N_{m+1} < i/N_{m+1} \leqslant j/N_m.$$

On multiplication through by N_{m+1} this gives the relations (1). *Therefore $A_{m+1}(\xi) \subseteq A_m(\xi)$ if $\xi \neq 1$*; and clearly also

$$A_{m+1}(1) \subseteq A_m(1).$$

For any ξ the compact sets $\overline{A_m(\xi)}$, $m = 1, 2, ...$, have thus at least one common point, and since $\Delta A_m(\xi) \to 0$, exactly one. This point is by definition $f(\xi)$.

f is continuous. If $|\xi - \eta| < 1/N_m$, $f(\xi)$ and $f(\eta)$ belong to the same or consecutive sets A_m^i, and therefore

$$\rho(f\xi, f\eta) < 2 \cdot 2^{-m+1}.$$

$S = f(<0, 1>)$. If $x \in A_m^{j+1}$, $\rho(x, f(j/N_m)) \leqslant \Delta(A_m^{j+1}) \leqslant 2^{-m+1}$. Hence the set of points $f(j/N_m)$, for all m and all j not exceeding N_m, is dense in S; and a fortiori the whole set $f(<0, 1>)$ is dense in S. But $f(<0, 1>)$ is closed and must therefore coincide with S.

Theorem 9·3. *In a locally compact, connected and locally connected space, S, every two points are connected by a locally connected continuum.*

Let a locally connected metric (7·1) be chosen in S and let ϵ_0 be as in Theorem 7·1. First let $a, b \in U(c, \epsilon)$, where $\overline{U(c, 2\epsilon)}$ is compact and $\epsilon < \epsilon_0$. Join a to b in the connected set $U(c, \epsilon)$ by a $\frac{1}{2}\epsilon$-chain, C_1. Join each consecutive pair of points, a_i, a_{i+1} of C_1 by a $\frac{1}{4}\epsilon$-chain in $U(a_i, \frac{1}{2}\epsilon)$. All these chains, combined in the order of the a_i, form a $\frac{1}{4}\epsilon$-chain, C_2, from a to b, lying in $U(c, 2\epsilon - \frac{1}{2}\epsilon)$. Proceeding in this way, we construct, for $m = 3, 4, \ldots$, an $\epsilon/2^m$-chain, C_m, from a to b, lying in $U(c, 2\epsilon - \epsilon/2^{m-1})$, each consecutive pair of points in C_{m-1} being joined by an $\epsilon/2^m$-chain of diameter $< \epsilon/2^{m-1}$ in C_m. Let $X = \overline{\cup C_m}$.

As a closed subset of $\overline{U(c, 2\epsilon)}$, X is compact, If $x, y \in X$, there are, for any m_0, points u, v of C_m within $\epsilon/2^{m_0}$ of x and y respectively, for some $m \geqslant m_0$; and u and v are $\epsilon/2^m$-connected in X. Thus x and y are $\epsilon/2^{m_0}$-connected in X for every m_0, and hence X is a connected set joining a to b. The same argument shews that, for any m, the $\epsilon/2^{m+1}$-chain joining a consecutive pair of points, a_i, a_{i+1} of C_m, together with the $\epsilon/2^{m+2}, \epsilon/2^{m+3}, \ldots$, chains derived from it, as described above, have as closure a compact connected set, Y_i, of diameter $\leqslant \epsilon/2^{m-1}$, containing a_i and a_{i+1}. The sets $\{Y_i\}$ are a finite covering of X by compact connected sets of diameter $\leqslant \epsilon/2^{m-1}$, and X is therefore locally connected.

Finally, if a and b are any points of S, let a connected chain of sets $V_1, V_2, \ldots, V_k$, with $a \in V_1$ and $b \in V_k$, be chosen from a covering of S by sets $U(x, \epsilon_x)$ such that $\overline{U(x, 2\epsilon_x)}$ is compact and $\epsilon_x < \epsilon_0$. If $a_0 = a, a_k = b$, and, for $0 < i < k, a_i \in V_i \cap V_{i+1}$, each pair of points a_i, a_{i+1} is joined by a locally connected continuum X_i; and $\cup X_i$ is the required set.

Theorem 9·3 shews that every two points of a locally connected continuum are the end-points of a *path*, i.e. a map of the segment $< 0, 1 >$ into the space. The stronger theorem can be proved that every two points are the end-points of a simple arc in the space.[13] For this reason the name *arc-wise connected* is given to spaces in which every two points are connected by a locally connected continuum.

§ 5. TOPOLOGICAL CHARACTERISATION OF
SEGMENT AND CIRCLE*

10. Among all particular topological spaces the most fundamental are the open line R^1 and its closed subset $<0, 1>$. It is therefore not very satisfactory that these spaces should be defined by means of the theory of real numbers, instead of by simple topological properties. In this section it will be shewn that the property of being disconnected by the removal of a single point may be used to define closed arcs (the homeomorphs of $<0, 1>$), and a similar characterisation is obtained for open arcs and simple closed curves. A characterisation on these lines was first given by Janiszewski.[14]

A *cut-point* of a connected set E is a point, x, such that $E-(x)$ is not connected.

Theorem 10·1. *If S is connected, but† $S-a$ has the partition $H_1 \mid H_2$, then $\bar{H}_1 = H_1 \cup a$ and $\bar{H}_2 = H_2 \cup a$.*

Since H_1 is closed in $S-a$,

$$H_1 = \bar{H}_1(S-a) = \bar{H}_1 - a,$$

and therefore $\bar{H}_1 \subseteq H_1 \cup a$. Similarly $\bar{H}_2 \subseteq H_2 \cup a$. If $\bar{H}_1 = H_1$, H_1 and $\bar{H}_2 \cup a$ are two non-null closed sets whose union is S and whose common part is

$$H_1(\bar{H}_2 \cup a) = H_1\bar{H}_2 \subseteq H_1(H_2 \cup a) = 0,$$

contrary to the assumption that S is connected.

It follows that $H_1 = S-\bar{H}_2$, $H_2 = S-\bar{H}_1$, and therefore H_1 and H_2 *are open sets* (in S).

Theorem 10·2. *A continuum, X, of which all but at most two points are cut-points is a simple arc.*

The compactness condition, implied in "continuum", is used not only to distinguish $<0, 1>$ from R^1 and its homeomorphs, but

---

* The contents of § 5 are used in this book only in proving one theorem: the converse of Jordan's Theorem (VI. 16·1 and 16·2). Apart, however, from its intrinsic importance the proof of 10·2 is a good extended example of the use of the methods of Part I.

† To simplify the formulae, $X-x$ and $X \cup a$ are written for $X-(x)$ and $X \cup (a)$ in this section.

also to exclude certain entirely different spaces which satisfy the cut-point condition. For example, the set of points in R^2 consisting of the portion $0 < \xi_1 \leqslant 1$ of the curve $\xi_2 = \sin 1/\xi_1$, together with the segment $< -1, 0 >$ of the ξ_1-axis, is connected and satisfies the cut-point condition; and the number of exceptional (non-cut) points is two, as it is in the simple arc; but the set is not locally connected, and therefore not the homeomorph of either $< 0, 1 >$ or R^1. The condition is still satisfied if any finite number of half-lines, $\xi_1 \leqslant 0$, parallel to the negative ξ_1-axis and lying between $\xi_2 = \pm 1$, are added to the set. This example, and also the example (Fig. 17) on p. 80, should be kept in mind in reading the first part of the following proof.

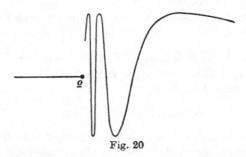

Fig. 20

Let X be the given space, and A the set of "exceptional points", i.e. points that are not cut-points. Thus A has 0, 1 or 2 members, and cannot be identical with X. Let x_0 be a point of X not in A, and $P \mid Q$ a partition of $X - x_0$.

(1) $\overline{P} = P \cup x_0$ and $\overline{Q} = Q \cup x_0$; and P and Q are open sets. This was proved in the preceding theorem.

(2) $\overline{P}$ and $\overline{Q}$ are connected. If $\overline{Q} = H_1 \mid H_2$, and $x_0 \in H_1$,

$$H_2 \overline{P} = H_2(P \cup x_0) = H_2 P = 0,$$

and therefore $X = H_2 \mid (H_1 \cup \overline{P})$.

(3) If $y \in P$ and $P_1 \mid Q_1$ is any partition of $X - y$, one of the parts P_1, Q_1 is contained in P. The connected set $\overline{Q}$ is contained in $X - y$ and therefore in one of the parts P_1, Q_1. The other part is contained in $P (= X - \overline{Q})$.

(4) P contains at least one exceptional point. Suppose it does not. Then $\overline{P} (= P \cup x_0)$ also contains no exceptional point. Let

x_1, x_2, ... be an enumerable dense set of points in P (II. 15·1, Corollary), and if $X - x_1 = P_1 \,|\, Q_1$, let P_1 be the part that is contained in P. We now make the inductive hypothesis that an integer n_r and a set P_r have been defined such that

$$X - x_{n_r} = P_r \,|\, Q_r$$

and $P_r \subseteq P$. Then P_r is a non-null open subset of P and therefore contains at least one of the points x_n. We define n_{r+1} to be the least integer such that $x_{n_{r+1}} \in P_r$. Since $P_r \subseteq P$, $x_{n_{r+1}}$ is not an exceptional point and therefore $X - x_{n_{r+1}}$ is not connected. Let $P_{r+1} \,|\, Q_{r+1}$ be a partition of it, P_{r+1} being the part contained in P_r (and therefore in P). This completes the inductive definition. It also shews, since $x_{n_{r+1}}$ belongs to P_r but not to P_{r+1}, that all the integers n_r are distinct.

The sequence of compact sets $\overline{P}$, $\overline{P}_1$, $\overline{P}_2$, ... is monotone and decreasing, and therefore the common part of all of them, P_∞, is not null. Since $\overline{P}_{s+1} = P_{s+1} \cup x_{n_{s+1}} \subseteq P_s$, P_∞ is also the common part of the sets P_s. Let z be any point of P_∞. Since $z \in P$, it is not an exceptional point. $X - z$ has therefore a partition $H_1 \,|\, H_2$, and by (3) every P_m contains H_1 or H_2. Therefore one of the parts, say H_1, is contained in P_m for an infinity of m, and therefore in P_∞. Since H_1 is open it contains a point, x_i, of the set (x_n). Let n_{s+1} be the first of the suffixes n_2, n_3, ... that exceeds i. Then $x_i \in P_s$, and n_{s+1} is not the least integer m such that $x_m \in P_s$, contrary to the definition in the previous paragraph.

This contradiction shews that the assumption that P contains no exceptional point is false.

It follows precisely similarly that Q contains at least one exceptional point, and therefore that there are at least two such points. We have thus proved that the number of exceptional points is two. We denote them by a and b.

(5) *If x is not an exceptional point $X - x$ has two components, each containing one of the two exceptional points.* Let $X - x = P \,|\, Q$, and let a belong to P and b to Q. We wish to shew that P and Q are connected sets. Suppose that $P = H_1 \,|\, H_2$ and that $a \in H_1$. Since H_2 is open and closed in P, which is open and closed in $X - x$, $X - x = H_2 \,|\, (X - x) - H_2$, which is impossible, since H_2 contains neither a nor b.

(*Note.* The properties (1) to (5) are all possessed by the set in Fig. 20. It will therefore certainly be necessary to make further use of the compactness of X.)

11. The results now established enable us to set up an *order* in X. A relation $\prec$ which holds between certain pairs of members of a set E is called a *total ordering* if it satisfies the conditions

(o_1) for no x is $x \prec x$;
(o_2) if $x \neq y$, either $x \prec y$ or $y \prec x$;
(o_3) if $x \prec y$ and $y \prec z$, then $x \prec z$.

(If o_2 is omitted $\prec$ is a *partial ordering*.) From o_1 and o_3 it follows that $x \prec y$ and $y \prec x$ are incompatible.

For any x of X we define L_x to be the null-set if $x = a$, and the component of $X - x$ containing a if $x \neq a$. R_x is similarly defined, b replacing a through-out. Thus for any x, $X = L_x \cup x \cup R_x$. No L_x contains b and no R_x contains a.

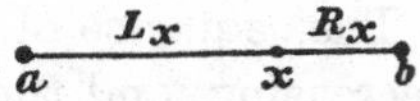

Fig. 21

(6) *The statements* $x \in L_y$ *and* $L_x \subset L_y$ *are equivalent.* First, if $x \in L_y$, $L_x \neq L_y$, since x belongs to L_y but not to L_x. The point y cannot be a since L_a is null; and if $y = b$, $L_x \subseteq X - b = L_b$. If $x = a$, $L_x = 0 \subseteq L_y$. Excluding all these cases, it follows from (3) that either $L_x \subseteq L_y$ or $R_x \subseteq L_y$; and since L_y does not contain b it must be $L_x \subseteq L_y$.

Secondly, given that $L_x \subset L_y$, again y cannot be a, and therefore if $x = a$, $x \in L_y$. Excluding this case, it follows from $L_x \subset L_y$ that

$$L_x \cup x = \bar{L}_x \subseteq \bar{L}_y = L_y \cup y.$$

But $x \neq y$ (since $L_x \neq L_y$) and therefore $x \in L_y$.

Similarly the statements $x \in R_y$ and $R_x \subset R_y$ are equivalent.

The relation $x \prec y$ is defined, for points of X, to mean "$x \in L_y$". Thus $a \prec x$ unless $x = a$, and $x \prec b$ unless $x = b$; and for no x does $x \prec a$ or $b \prec x$. (The symbol $x \prec y$ may be read "x precedes y", and such expressions as "first point", "between", "successor" will be used accordingly.)

(7) *The relation* $\prec$ *is a total ordering.* o_1 is obviously satisfied. o_2: If x does not precede y then $x \in (X - L_y) = R_y \cup y$; and if also $x \neq y$, $x \in R_y$. Hence $R_x \subseteq R_y$, and, taking complements,

$L_y \cup y \subseteq L_x \cup x$. Since $x \neq y$, it follows that $y \in L_x$, i.e. $y \prec x$. o_3 follows immediately from (6).

The set $[x \prec p]$, for any fixed p, is L_p itself, and therefore an open set; $[p \prec x]$ is, by the conditions o, the complement of $[x \prec p]$ in $X - p$, i.e. it is R_p, and therefore also open. Hence the set $[p \prec x \prec q]$, which is the common part of two such sets, is open. We denote it by $\prec p, q \succ$.

(8) *If $p \prec q$ the set $\prec p, q \succ$ is not null.* For if it were, every point of X would belong either to $L_p \cup p$ or to $R_q \cup q$. These closed sets have no common point, for such a point would belong both to L_p and to R_q, since $p \prec q$; i.e. it would precede p and follow q, which, by o_3, is impossible. Thus if $\prec p, q \succ$ were null,

$$(L_p \cup p) \mid (R_q \cup q)$$

would be a partition of X.

Let E_0 be an enumerable dense set in X, not containing a or b, say
$$E_0 = (x_1, x_2, \ldots).$$

If $p \prec q$, the non-null open set $\prec p, q \succ$ contains at least one point of E_0, and in particular there is a point of E_0 between any two points of E_0 itself.

Let α_1, α_2, ... be any enumeration of the rational points of $(0, 1)$. We construct two sequences

$y_1, y_2, \ldots$ of points of E_0,
$\beta_1, \beta_2, \ldots$ of rational points of $(0, 1)$,

as follows. Let y_1 be x_1 and β_1 be α_1. Suppose that y_r and β_r have been defined for $r = 1, 2, \ldots, n-1$. If n is even, let y_n be the point x_k of lowest k not already in the set $y_1, y_2, \ldots, y_{n-1}$, and let β_n be the α_k of lowest k having the same relations to $\beta_1, \beta_2, \ldots, \beta_{n-1}$, relative to $<$, as y_n has to $y_1, y_2, \ldots, y_{n-1}$, relative to $\prec$. If n is odd the roles of x's and α's, y's and β's, are reversed: β_n is the number α_k of lowest k not among the set $\beta_1, \beta_2, \ldots, \beta_{n-1}$, and y_n the x_k of lowest k with the right $\prec$-relations. (That such a point exists follows from (8).)

Clearly every x_i appears once, and only once, as a y_j, and every α_i as a β_j. Thus if $f(y_i)$ is defined to be β_i, f is a $(1, 1)$ order-preserving mapping of E_0 on to the rational points of $(0, 1)$.

A subset, Λ, of E_0 is a *section* if

  (i) it has no last point;

  (ii) if $x \in \Lambda$ all predecessors of x in E_0 belong to Λ.

(Λ may be null or the whole of E_0.) The analogy with sections of the rational numbers in $(0, 1)$ is obvious: it is made exact by the following result, the analogue of Dedekind's Theorem.

  (9) *Let Λ be a section of E_0, and K the set of points of X not followed by any point of Λ. Then K has a first point.* K cannot be null for it contains b. If $K = X$ the required first point is a; we therefore exclude this case. $X - K$ is an open set, for if x is one of its points, and y is a point of Λ following x, $\prec a, y \succ$ is a neighbourhood of x (i.e. an open set containing x) in $X - K$. If K has no first point K is also open, for if $x \in K$ there is a point, y, which precedes x and follows all points of Λ; and $\prec y, b \succ$ is a neighbourhood of x in K. Thus the supposition that K has no first point leads to the conclusion that $K \mid (X - K)$ is a partition of X.

It has now been shewn that to every section of E_0 corresponds a unique point of X, which we call the point determined by the section. The points determined by two different sections are different, for if x belongs to Λ_1 but not to Λ_2, x precedes the point determined by Λ_1 but not that determined by Λ_2. Every point, x, of X is determined by at least one section, namely the set of all points of E_0 that precede x. We have therefore set up a $(1, 1)$-correspondence between the sections of E_0 and the points of X.

The mapping f of E_0 on the rational points of $(0, 1)$ may now be extended to all points of X. If x is any point of X, and Λ_x the section determining it, f, since it is order-preserving, clearly maps Λ_x on a section of the rationals of $(0, 1)$. We define $f(x)$ to be the real number determined by this section. From what has now been proved it is clear that f is a $(1, 1)$-mapping of X on $< 0, 1 >$, and that for members of E_0 the new definition of $f(x)$ agrees with the old.

  (10) *The mapping f, so extended, is order-preserving*, i.e. if $x \prec y$ then $f(x) < f(y)$. If x and y both belong to E_0 this follows from the original definition of f. If $x \in E_0$ but y does not, $x \in \Lambda_y$, and therefore $f(x) \in f(\Lambda_y)$, the section of the rationals determining

$f(y)$. Hence $f(x) < f(y)$. If $y \in E_0$ but x does not, y does not belong
to Λ_x, and therefore $f(y)$ does not belong to $f(\Lambda_x)$, i.e. $f(x) < f(y)$.
From these two cases, and the fact that between any two points of
X there is a point of E_0, the general case now follows, on using o_3.

(11) *f is a topological mapping of X on $< 0, 1 >$.* First, f is
continuous. To prove this it is sufficient (by III. 9·2) to shew
that if G is an open set in $< 0, 1 >$, $f^{-1}(G)$ is open in X; and for this
it is sufficient to shew that the open intervals (open in $< 0, 1 >$!)
which are the components of G are mapped on open sets in X.
But the typical open intervals

$$(\gamma, \delta), \quad (\gamma, 1 >, \quad < 0, \gamma)$$

in $< 0, 1 >$ are mapped on $\prec c, d \succ$, $[c \prec x]$, and $[x \prec c]$ (where
$f(c) = \gamma$ and $f(d) = \delta$); and these are open sets.

Thus f is continuous, and therefore, since X is compact, it is
a homeomorphism. (This is the predicted further use of the
compactness condition.)

The theorem is thus completely proved.

Corollary 1. *a and b are the end-points.*

Theorem 11·1. *Every continuum has at least two points that are
not cut-points.* (Proved in para. 10.)

Exercise. At what point does an attempt to avoid the use of compactness at the end of the proof by proving directly that f^{-1} is
continuous break down?

The *open arc* (homeomorph of R^1) may now be defined in-
trinsically as the homeomorph of a closed arc without its end-
points. This characterisation, although genuinely topological, is
too indirect to be entirely satisfactory. The open arc is also
characterised as a space, Y, with the following combination of
properties:

(1) separable;
(2) connected and locally connected;
(3) if x is any point, $Y - x$ has two components.[15]

12. Theorem 12·1. *A continuum whose connection is destroyed
by the removal of two arbitrary points is a simple closed curve.*

Let x and y be two points that are not cut-points of the space
S (Theorem 11·1 above). Let $S - (x \cup y) = P_1 \,|\, P_2$.

(1) $\overline{P}_i = P_i \cup x \cup y$ $(i = 1, 2)$. For $\overline{P}_i \subseteq P_i \cup x \cup y$ as in the proof of 10·1. Since $S - x$ is connected and y a cut-point of it, $y \in \overline{P}_1$ and $y \in \overline{P}_2$ by 10·1; and similarly $x \in \overline{P}_1$ and $x \in \overline{P}_2$.

(2) $\overline{P}_i$ *is connected.* Suppose first that $P_1 \cup x = H_1 \mid H_2$, where $x \in H_1$. The sets $P_i \cup x = \overline{P}_i - y$ are closed in $S - y$, hence H_1 and H_2 are also closed. Therefore

$$S - y = H_2 \mid H_1 \cup (P_2 \cup x)$$

is a partition, contrary to the assumption that y is not a cut-point of S. Thus $P_1 \cup x$ is connected, hence also $\overline{P}_1 = \overline{P_1 \cup x}$, and similarly $\overline{P}_2$.

(3) *Every point of P_i is a cut-point of $\overline{P}_i$.* Suppose, for example, that $\overline{P}_1 - u$ is connected, where $u \in P_1$ ("assumption A"). Then if $v \in P_2$, $\overline{P}_2 - v$ cannot also be connected for otherwise

$$(\overline{P}_1 - u) \cup (\overline{P}_2 - v),$$

i.e. $S - (u \cup v)$, would be connected, contrary to the data. It follows from 10·2 that $\overline{P}_2$ is a simple arc, with end-points x and y. Hence if $v \in P_2$, $\overline{P}_2 - v$ has two components, C_x and C_y, containing x and y respectively; and therefore (on assumption A)

$$C_x \cup (\overline{P}_1 - u) \cup C_y,$$

i.e. $S - (u \cup v)$, is connected. This contradiction shews that assumption A is false.

From (1), (2) and (3) it follows that $\overline{P}_i$ is a simple arc with end-points x and y. Hence S is the union of two arcs with the same end-points but no other common point, i.e. a simple closed curve.

Chapter V

SEPARATION THEOREMS

§ 1. CHAINS ON A GRATING

1. It is convenient in the following chapters to use X^2 to denote either R^2 (the open plane) or Z^2 (the closed plane). In any one argument X^2 denotes, of course, one definite space.

A *rectangular grating*, **G**, in the open or closed plane, is formed by drawing a finite number of segments across a square, parallel to its sides. The lines go right across the square: divisions like that in Fig. 22(*b*) are not allowed. The grating in which the square is left undivided is not excluded.

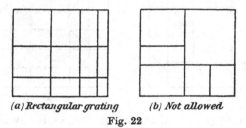

(a) *Rectangular grating*   (b) *Not allowed*

Fig. 22

For convenience we suppose that the original square (the *frame* of the grating) has its centre at the origin and its sides parallel to the coordinate axes, so that the words "horizontal", "above", "below", etc. may be used in an obvious sense.

The 2-*cells* of **G** are (1) the closures of the rectangular domains into which the interior of the frame is divided, and (2) the closure of the exterior of the frame (including the point at infinity if there is one). The *edges*, or 1-*cells*, are the sides of the 2-cells of finite diameter, and the *vertices*, or 0-*cells*, are their corners.

Thus all cells are bounded closed sets of points, except the 2-cell outside the frame, which is closed but not bounded.†

Every edge of the grating evidently belongs to just two 2-cells, which lie on opposite sides of it.

† To avoid confusion with another sense of "bounded", used in this chapter, this will be called the *infinite* 2-cell; and all others *finite* cells.

Theorem 1·1. *If F_1 and F_2 are non-intersecting closed sets in Z^2, there exists a grating,* **G,** *no cell of which meets both F_1 and F_2.*

At least one of the closed sets, say F_1, does not contain w, and therefore lies, for some α, in the interior of the set

$$E_\alpha : \quad [\,|\,\xi_1\,| \leqslant \alpha, |\,\xi_2\,| \leqslant \alpha].$$

The non-intersecting compact sets $F_2 E_\alpha$ and F_1 in R^2 are at a positive distance δ apart. Let m be an integer exceeding $4\alpha/\delta$. If the frontier of E_α is taken as frame, and m equidistant segments are drawn between, and parallel to, each pair of sides the result is evidently a grating with the required property.

In R^2 the additional condition that *at least one of the closed sets is compact* is required. The proof is then as before.

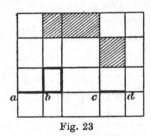

Fig. 23

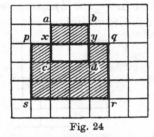

Fig. 24

2. A *k-chain,* C^k ($k = 0, 1, 2$), on a grating **G,** is any set of k-cells of **G.** The *sum* (modulo 2) of two k-chains, C_1^k and C_2^k, is the set of k-cells that belong to one, but not both, of C_1^k and C_2^k. It is denoted by $C_1^k + C_2^k$. The complement $\mathscr{C}C^2$ of a 2-chain C^2 is the set of 2-cells of **G** not belonging to C^2. Thus if Ω^2 denotes the 2-chain containing all the 2-cells of **G,** $\mathscr{C}C^2 = C^2 + \Omega^2$.

Example. 1. In Fig. 23 the set of three shaded cells is a 2-chain, and the six thick edges form a 1-chain. In Fig. 24, if C_1^2 and C_2^2 are the 2-chains whose cells are those inside the rectangles *abdc* and *pqrs,* $C_1^2 + C_2^2$ is the set of shaded 2-cells. $\mathscr{C}C_1^2 + \mathscr{C}C_2^2$ is also the set of shaded 2-cells.

Addition of k-chains is commutative and associative, and

$$\sum_1^q C_i^k = C_1^k + C_2^k + \dots + C_q^k$$

is the set of k-cells that belong to an odd number of the chains C_i^k. Clearly for any C^k, $C^k + C^k = 0$ (the "zero chain"), and hence the equation $C_1^k + X^k = C_2^k$ is satisfied by $X^k = C_1^k + C_2^k$ and no other k-chain. Thus the k-chains on $\mathbf{G}$ form a commutative group under the operation of addition modulo 2. Some of the consequences of this fact are considered in the final section of Chapter VI, but otherwise no use is made of the language or theorems of group-theory.

The *boundary*, $\overset{\bullet}{C}{}^k$, of the k-chain C^k on $\mathbf{G}$ is (for $k = 1, 2$) the set of ($k-1$)-cells of $\mathbf{G}$ that are contained in an odd number of k-cells of C^k. (The boundary of a 0-chain is not defined.)

Example. 2. The boundary of an edge is its pair of end-points. The boundary of the 1-chain in Fig. 23 is the set of vertices a, b, c, d. The boundary of the shaded 2-chain in Fig. 24 is the set of the thickened edges.

Theorem 2·1. $(C_1^k + C_2^k)^{\bullet} = \overset{\bullet}{C}{}_1^k + \overset{\bullet}{C}{}_2^k$ $(k = 1, 2)$.

A $(k-1)$-cell A^{k-1} belonging to n_1 k-cells of C_1^k and n_2 of C_2^k belongs to $\overset{\bullet}{C}{}_i^k$ if n_i is odd, and therefore to

$$\overset{\bullet}{C}{}_1^k + \overset{\bullet}{C}{}_2^k$$

if $n_1 + n_2$ is odd. This is the condition that it belongs to $(C_1^k + C_2^k)^{\bullet}$.

By induction on the number of chains it follows that

$$(C_1^k + C_2^k + \ldots + C_q^k)^{\bullet} = \overset{\bullet}{C}{}_1^k + \overset{\bullet}{C}{}_2^k + \ldots + \overset{\bullet}{C}{}_q^k.$$

In particular the boundary of any k-chain is the sum, mod 2, of the boundaries of its k-cells.

Since $\overset{\bullet}{\Omega}{}^2 = 0$ it follows from 2·1 that, for any C^2,

$$(\mathscr{C}C^2)^{\bullet} = (C^2 + \Omega^2)^{\bullet} = \overset{\bullet}{C}{}^2.$$

A *k-cycle* is, for $k = 1, 2$, a k-chain whose boundary is zero; a *0-cycle* is a 0-chain with an even number of 0-cells. The sum mod 2 of any set of k-cycles is a k-cycle (by 2·1, or directly for $k = 0$).

Theorem 2·2. *The boundary of any k-chain is a $(k-1)$-cycle* $(k = 1, 2)$.

The boundary of a 1-cell is two vertices—a 0-cycle. The boundary of a 2-cell is a 1-chain in which every vertex belongs to just two edges, i.e. it is a 1-cycle. From these special cases the theorem follows by addition.

3. The k-chain C^k (a finite set whose members are k-cells) is to be distinguished from the union of its k-cells, a set of points denoted by $|C^k|$ and called the *locus* of C^k, or the *set of points covered by* C^k. The functions, $\mathscr{F}$, $\mathscr{K}$ (or $^-$) and $\mathscr{I}$ are applicable to the locus $|C^k|$ but not, of course, to C^k itself. To avoid clumsy constructions, a chain (instead of, more correctly, the locus of a chain) will often be said to contain a point, to lie in a set of points, or to meet a set or another chain; but this licence will not extend to symbolic relations.†

Clearly $|C_1^k + C_2^k| \subseteq |C_1^k| \cup |C_2^k|$ in all cases. The two sets are identical if, and only if, C_1^k and C_2^k have no common k-cell. (Thus in Example 1 above, $|C_1^k| \cup |C_2^k|$ is the whole closed region bounded by the polygon $abyqrspx$, $|C_1^k + C_2^k|$ is the shaded region.) Note that whereas $\mathscr{C}|C^2|$ is an open set, $|\mathscr{C}C^2|$ is closed, and is in fact $\overline{\mathscr{C}|C^2|}$.

A k-chain C^k is, by definition, *connected* if its locus $|C^k|$ is connected. The maximal connected k-chains contained in any k-chain C^k are called the *components* of C^k. They have as loci the components of $|C^k|$.

Theorem 3·1. *If K^k is a component of C^k, $\mathring{K}^k$ is the part of $\mathring{C}^k$ in K^k ($k > 0$).*

For if a cell A^{k-1} is contained in a k-cell of K^k, all the cells of C^k containing A^{k-1} are in K^k, and therefore A^{k-1} has the same parity in C^k and K^k.

The following simple deduction from 3·1 is one of the main links between the combinatorial theory of this section and the point-set notions of Part I.

Theorem 3·2. *If x and y form the boundary of a 1-chain C^1, they are connected in $|C^1|$.*

If not, the component of C^1 containing x has, by 3·1, the single odd vertex x as its boundary, contrary to 2·2.

Theorem 3·3. *For any C^2, $\mathscr{F}|C^2| = |\mathring{C}^2|$.*
If $x \in |\mathring{C}^2|$ then

$$x \in |C^2| \cap |\mathscr{C}C^2| = |C^2| \cap \overline{\mathscr{C}|C^2|} = \mathscr{F}|C^2|.$$

† Thus we shall write "x lies in C^k", but "$x \in |C^k|$".

If $x \in \mathscr{F} \mid C^2 \mid$, at least one 2-cell containing x is in C^2, and one is in $\mathscr{C}C^2$. Therefore a 2-cell of C^2 and one of $\mathscr{C}C^2$ have an edge in common that contains x. Thus $x \in \mid \mathring{C}^2 \mid$.

It follows from 3·3 that Ω^2 and 0 are the only 2-cycles. Hence if $\mathring{C}^2$ is connected so also is C^2, for either $C^2 = \Omega^2$ or else every component of C^2 contains a non-null component of $\mathring{C}^2$.

4. It is sometimes necessary to pass from one grating, **G**, to another, **G***, by introducing additional lines; or by enlarging the frame, i.e. by extending all the lines to meet a larger square containing the original frame; or by a combination of both processes. Such a new grating **G*** is called a *refinement* of the old. (It is convenient to agree that **G** is a refinement of itself.) A common refinement can be formed for any two gratings **G₁** and **G₂**, by taking as frame any square, with sides parallel to the axes, containing the frames of **G₁** and **G₂**, and as cross-lines all the lines of **G₁** and **G₂**, suitably extended.

To each k-chain, C^k, on **G** corresponds the *subdivided* k-chain C^k_* on **G***, which is the sum of the k-chains into which the k-cells of C^k are subdivided. (0-chains are unaltered by subdivision: $C^0_* = C^0$.) A subdivided chain has the same locus as its original: $\mid C^k_* \mid = \mid C^k \mid$.

Theorem 4·1. $(C^k_1 + C^k_2)_* = C^k_{1*} + C^k_{2*}$, $(\mathring{C}^k)_* = (C^k_*)^\bullet$.

The first relation is clear from the definitions.

The effect of subdivision on an edge is to introduce new intermediate vertices, but since they each belong to two new edges the boundary of the resulting 1-chain is still the original two endpoints. Hence, by addition, $(C^1_*)^\bullet = \mathring{C}^1 = (\mathring{C}^1)_*$. If C^2 is any 2-chain,

$$\mid (C^2_*)^\bullet \mid = \mathscr{F} \mid C^2_* \mid = \mathscr{F} \mid C^2 \mid = \mid \mathring{C}^2 \mid = \mid (\mathring{C}^2)_* \mid.$$

Thus $(C^2_*)^\bullet$ and $(\mathring{C}^2)_*$ are chains on **G*** with the same locus, and therefore identical.

Corollary. C^k_* *is a k-cycle if, and only if, C^k is a k-cycle.*

5. Theorem 5·1. *Every 1-cycle on a rectangular grating in Z^2 or R^2 is the boundary of just two 2-chains.*

This theorem is called the Fundamental Lemma because, in

spite of its simplicity, it is the geometrical kernel of the whole of the theory in this Chapter. It is a property of the plane and sphere, and distinguishes them topologically from all other surfaces. For example, if the surface of a solid ring is divided into squares, a meridian circle, though a "1-cycle", is not the boundary of any 2-chain on the surface (Fig. 25).

The proof is by induction on the number of lines drawn across the original square to make the grating. On the grating consisting of the frame alone, the only 1-cycles are the null-set, which bounds two 2-chains (the zero-chain and Ω^2); and the square itself, which bounds the two 2-cells possessed by this grating.

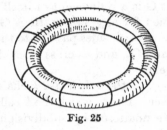

Fig. 25

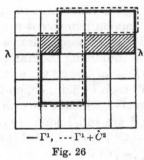

$-\Gamma^1, \cdots \Gamma^1 + \dot{C}^2$

Fig. 26

Let the given grating, G_1, be formed from a grating G_0, for which the theorem may be assumed true, by the addition of a line λ across the square, which we may suppose to be parallel to the ξ_1-axis. Let Γ^1 be the given 1-cycle on G_1. We denote by C^2 the sum of the 2-cells of G_1 whose lower edges lie in the line λ and belong to Γ^1. The 1-cycle $\Gamma^1 + \dot{C}^2$ therefore contains no edge in λ. *It follows that it is the subdivided form of a 1-cycle Γ_0^1 on G_0,* for $\Gamma^1 + \dot{C}^2$ contains no horizontal edge at a vertex, x, of λ, and therefore, since it is a cycle, contains both or neither of the vertical edges at x, which together make up an edge of G_0.

By hypothesis there is a 2-chain C_0^2 on G_0 such that $\Gamma_0^1 = \dot{C}_0^2$, and if C_1^2 is the subdivided form of C_0^2 on G_1, $\dot{C}_1^2 = \Gamma^1 + \dot{C}^2$. Hence

$$(C_1^2 + C^2)^\bullet = (\Gamma^1 + \dot{C}^2) + \dot{C}^2 = \Gamma^1.$$

The residual 2-chain $\mathscr{C}(C_1^2 + C^2)$ also has boundary Γ^1.

It has thus been shewn that Γ^1 bounds at least two 2-chains. *No 1-cycle bounds more than two 2-chains.* For if $\overset{\bullet}{C}{}^2_a = \overset{\bullet}{C}{}^2_b$, $(C^2_a + C^2_b)^\bullet = 0$ and therefore $C^2_a + C^2_b$ is 0 or Ω^2, i.e. $C^2_b = C^2_a$ or $\mathscr{C}C^2_a$.

6. The convention is now introduced that if C^k is a k-chain on a grating **G**, the corresponding subdivided chain on any refinement **G*** shall be denoted by the same symbol C^k, with the addition when necessary of the words "on **G***". Thus if C^k_1 and C^k_2 are chains on the gratings $\mathbf{G_1}$ and $\mathbf{G_2}$, and **G*** is any common refinement of $\mathbf{G_1}$ and $\mathbf{G_2}$, "$C^k_1 + C^k_2$ (on **G***)" will denote $C^k_{1*} + C^k_{2*}$. No ambiguity can arise from the various orders in which a combination of subdivision, addition, and taking the boundary can be carried out, since, by 2·1 and 4·1 the result is always the same.

Let G be an open set and Γ^k a k-cycle on **G** in G. The cycle Γ^k *bounds in* G ($\Gamma^k \sim 0$ in G) if there is, on some refinement **G*** of **G**, a $(k+1)$-chain C^{k+1} such that $|C^{k+1}| \subseteq G$ and $\overset{\bullet}{C}{}^{k+1} = \Gamma^k$ on **G***. "Γ^k is non-bounding in G" means that $|\Gamma^k| \subseteq G$ but Γ^k does not bound in G.

Examples. 1. If A^2 is a 2-cell of **G**, $\overset{\bullet}{A}{}^2 \sim 0$ in X^2 on **G** itself; but if $a \in \mathscr{I} \mid A^2 \mid$ and $b \in \mathscr{C} \mid A^2 \mid$, $\overset{\bullet}{A}{}^2$ does not bound in $X^2 - (a \cup b)$ because (as will be shewn below) every 2-chain bounded by $\overset{\bullet}{A}{}^2$ contains a or b.

2. If x and y are finite points of the plane, $(x) + (y) \sim 0$ in X^2, for if **G** is a grating having x and y as vertices, $(x) + (y)$ bounds either an edge or an L-shaped 1-chain on **G**.

It is often unnecessary to make explicit reference to a grating in the enunciation of theorems. Thus the opening phrase of the following theorem (6·1) means "If Γ^k_i is a k-cycle in G on a grating $\mathbf{G}_i$," or still more explicitly "If Γ^k_i is a k-cycle on $\mathbf{G}_i$ and $\mid \Gamma^k_i \mid \subseteq G$". Similarly if x and y are any points, "$(x) + (y) \sim 0$ in G" means "$(x) + (y)$, regarded as a 0-cycle on some grating **G** having x and y as vertices, bounds in G".

Theorem 6·1. *If Γ^k_i is a cycle in G and $\Gamma^k_i \sim 0$ in G, for*

$$i = 1, 2, \ldots, q,$$

then $\sum\limits_1^q \Gamma^k_i \sim 0$ in G.

If $\Gamma_i^k = \dot{C}_i^{k+1}$ on $\mathbf{G}_i^*$, and if $\mathbf{G}_0$ is a common refinement of all the $\mathbf{G}_i^*$, then

$$\sum_1^q \Gamma_i^k = \left(\sum_1^q C_i^{k+1}\right)^{\bullet}$$

on $\mathbf{G}_0$.

Corollary. *If Γ^k is non-bounding in G, at least one of its components is non-bounding in G,* for since the components do not meet each other their (mod 2) sum is Γ^k.

Theorem 6·2. *If $\mathscr{C}G$ is connected every 1-cycle in G bounds in G.*

For if $\mathscr{C}G$ met both the 2-chains bounded by Γ^1 it would meet the common frontier, $|\,\Gamma^1\,|$, of their loci.

The relation $\Gamma_1^k + \Gamma_2^k \sim 0$ in G is also written $\Gamma_1^k \sim \Gamma_2^k$ in G, and read "Γ_1^k is homologous to Γ_2^k in G". Homology in G is an equivalence relation, for it is clearly symmetrical; it is reflexive since $\Gamma^k + \Gamma^k = 0$; and it is transitive, for if $\Gamma_1^k \sim \Gamma_2^k$ and $\Gamma_2^k \sim \Gamma_3^k$ in G, then
$$\Gamma_1^k + \Gamma_3^k = (\Gamma_1^k + \Gamma_2^k) + (\Gamma_2^k + \Gamma_3^k) \sim 0 \text{ in } G.$$

Exercise. State and prove 6·1 without making use of the convention introduced at the beginning of the paragraph.

Theorem 6·3. *A necessary and sufficient condition that the finite points x and y be connected in an open set G is that $(x) + (y) \sim 0$ in G.*

Necessary. By hypothesis x and y belong to the same component, D, of G. Let E_x be the set of all points z such that $(x) \sim (z)$ in D. E_x is open, for if $z \in E_x$ every point z' in a sufficiently small neighbourhood of z is connected to z by either one or two segments in D. The sum, mod 2, of this chain and the one which (by hypothesis) is bounded by x and z is a 1-chain in D bounded by x and z', i.e. $z' \in E_x$. $D - E_x$ *is open,* for, as before, if x and a

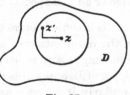

Fig. 27

point z', in a small neighbourhood of z of $D - E_x$, were together the boundary of a 1-chain in D it would follow that x and z bound in D, contrary to the definition of $D - E_x$. Since D is connected one of the sets E_x and $D - E_x$ must be null; and since $x \in E_x$, $D - E_x = 0$. Therefore $y \in E_x$.

Sufficient. This follows immediately from 3·2.

Theorem 6·4. *If x and y do not lie on $|\Gamma_1^1|$ or $|\Gamma_2^1|$, at least one of the 1-cycles Γ_1^1, Γ_2^1, $\Gamma_1^1 + \Gamma_2^1$ bounds in $X^2 - (x \cup y)$.*

Let C_i^2 be the 2-chain, on $\mathbf{G}_i$ say, that contains x and is bounded by Γ_i^1 ($i = 1, 2$). If, for $i = 1$ or 2, y is also in $|C_i^2|$ then $\Gamma_i^1 = (\mathscr{C}C_i^2)^{\bullet} \sim 0$ in $X^2 - (x \cup y)$. If y is in neither $|C_1^2|$ nor $|C_2^2|$ it is not in $|C_1^2 + C_2^2|$. Since every cell containing x belongs to both C_1^2 and C_2^2, x is not in $|C_1^2 + C_2^2|$. Thus

$$\Gamma_1^1 + \Gamma_2^1 = (C_1^2 + C_2^2)^{\bullet} \sim 0 \text{ in } X^2 - (x \cup y).$$

7. *Change of notation.* It has so far been possible to discuss the properties of 0-, 1- and 2-chains to some extent together, but from now on their parts in the theory are different. To avoid a multiplicity of suffixes it is convenient to adopt new notations. We use

A for a 2-cell, K for a 2-chain;

α for a 1-cell, γ for a 1-cycle, κ for a general 1-chain;

italic letters x, y, a, b, etc. for vertices.

General 0-chains are not often referred to, and are then denoted, as before, by C^0, 0-cycles by Γ^0.

The 0-chain (x) is always to have its brackets, e.g. in $(x) \sim (y)$. This prevents the possibility of confusion between the 0-chain $(x) + (y)$ and the vector sum of points $x + y$.

8. Two 1-chains, κ_1 and κ_2, on a grating have *general intersection* if they cross at every common vertex, that is to say if (i) no common vertex lies on the outer frame of the grating, and (ii) at each common vertex the two edges parallel to $\xi_1 = 0$ belong to one 1-chain, and the other two edges to the other.

Theorem 8·1. *If a 1-cycle γ and a 1-chain κ with boundary $(x) + (y)$ have general intersection, $\gamma \sim 0$ in $X^2 - (x \cup y)$ if, and only if, the number, n, of crossings is even.*

Let κ_1 and κ_2 be the subchains of κ in K_1 and K_2, the 2-chains bounded by γ. Each common vertex of κ and γ belongs to just one edge of κ_1, and therefore $\dot{\kappa}_1$ contains all these vertices. If n is odd, the 0-cycle $\dot{\kappa}_1$ cannot consist precisely of this odd set of vertices, and therefore κ_1 contains one of the vertices x, y—the only other available odd vertices—and κ_2 contains the other. If

n is even, $\overset{\bullet}{\kappa}_1$ contains both or neither of x and y, and κ_2 neither or both, accordingly.

Corollary. *Two 1-cycles, γ_1 and γ_2, with general intersection have an even number of crossings.*

Theorem 8·2. *If, for $i = 1$, 2, κ_i is a 1-chain with boundary $(x_i) + (y_i)$; if κ_1 and κ_2 have general intersection, with an odd number of crossings; and if the continuum F contains x_1 and y_1 but does not meet κ_2; then $F \cup |\kappa_1|$ separates x_2 and y_2.*

If not let κ_3 be a 1-chain in $\mathscr{C}(F \cup |\kappa_1|)$ with boundary $(x_2) + (y_2)$. The 1-cycle $\kappa_2 + \kappa_3$ has general intersection with κ_1, the crossings being the same as those of κ_1 and κ_2 and therefore odd in number. Therefore x_1 and y_1 belong to different 2-chains bounded by $\kappa_2 + \kappa_3$. But this is impossible, for they are joined by the continuum F, meeting neither κ_2 nor κ_3.

These theorems can be made the basis of alternative proofs of the separation theorems that occupy the rest of this chapter. More general forms of 8·1 and 8·2 are proved in Chapter VII (8·5 Corollary).

§ 2. ALEXANDER'S LEMMA AND JORDAN'S THEOREM

9. A theorem ("Alexander's Lemma")[16] will now be proved which gives an answer to a question that often arises in the course of this chapter. Suppose it is given that two points can be joined by a path not meeting a closed set F_1, and also by a path not meeting a closed set F_2; in what circumstances can they be joined by a path meeting neither F_1 nor F_2? Or, in terms of open sets, given that $(x) \sim (y)$ in G_1 and in G_2, in what circumstances does $(x) \sim (y)$ in $G_1 G_2$?

The lemma can be stated in two ways, which are easily seen to be equivalent, by taking G_1 and G_2 in the first version to be the complements of F_1 and F_2 in the second.

Theorem 9·1·1. *Let G_1 and G_2 be open sets in the closed plane Z^2. If the points x and y bound the 1-chains κ_1 in G_1 and κ_2 in G_2, and if $\kappa_1 + \kappa_2 \sim 0$ in $G_1 \cup G_2$, then $(x) \sim (y)$ in $G_1 G_2$.*

Theorem 9·1·2. *Let F_1 and F_2 be closed sets in the closed plane Z^2. If the points x and y bound the 1-chains κ_i not meeting F_i (for*

$i = 1, 2$) *and if* $\kappa_1 + \kappa_2 \sim 0$ *in* $Z^2 - F_1 F_2$, *then* x *and* y *are not separated by* $F_1 \cup F_2$.

We shall prove the second version, 9·1·2, which, in spite of its less direct form, is more readily applicable in most cases.

By hypothesis, one of the 2-chains bounded by $\kappa_1 + \kappa_2$, say K on a grating $\mathbf{G}$, does not meet $F_1 F_2$, and hence $|K|F_1$ and $|K|F_2$ do not meet. Let $\mathbf{G^*}$ be a refinement of $\mathbf{G}$ such that no cell meets both $|K|F_1$ and $|K|F_2$, and let K_1 be the set of 2-cells of K^* ($= K$ as subdivided on $\mathbf{G^*}$) that meet F_1. Consider the 1-chain $\kappa_0 = \kappa_2 + \overset{*}{K_1}$ on $\mathbf{G^*}$. (In Fig. 28 K_1 is horizontally shaded, and κ_0 is the thickened lines.) Since $\overset{*}{K_1}$ is a cycle $\overset{*}{\kappa_0} = \overset{*}{\kappa_2}$, i.e. κ_0 is bounded by x and y.

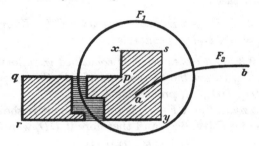

Fig. 28

κ_0 *does not meet* $F_1 \cup F_2$. By the definition of κ_2, $|\kappa_2| \cap F_2 = 0$, and since all cells of K_1 meet F_1, $|K_1|F_2 = 0$. Therefore κ_0 does not meet F_2. The set F_1 does not meet $K + K_1$, nor a fortiori its boundary $\kappa_1 + \kappa_2 + \overset{*}{K_1}$. Since $|\kappa_1|F_1 = 0$, and

$$|\kappa_0| = |\kappa_2 + \overset{*}{K_1}| = |\kappa_1 + (\kappa_1 + \kappa_2 + \overset{*}{K_1})| \subseteq |\kappa_1| \cup |\kappa_1 + \kappa_2 + \overset{*}{K_1}|,$$

it follows that κ_0 does not meet F_1.

This completes the proof.

Corollary. *In* 9·1·1 *the vertex-pair* $(x) + (y)$ *may be replaced by any* 0*-cycle* Γ^0 *bounding* κ_1 *in* G_1 *and* κ_2 *in* G_2. The proof is unaffected.

In the *open plane* the condition "F_1 or F_2 is bounded" must be added, to ensure the existence of the grating $\mathbf{G^*}$; but this extra condition is not necessary in the case $F_1 F_2 = 0$, for then K can always be chosen to be finite.

Theorem 9·2. *If the common part of the two closed sets F_1 and F_2 in Z^2 is connected, two points which are connected in $\mathscr{C}F_1$ and $\mathscr{C}F_2$ are connected in $\mathscr{C}(F_1 \cup F_2)$.* (Holds in R^2 if $F_1 F_2 = 0$.)

By a preliminary topological mapping of the whole space, which does not disturb separation properties, the points may both be made finite. They then bound chains κ_i not meeting F_i, and since $F_1 F_2$ is connected $\kappa_1 + \kappa_2$ bounds in $\mathscr{C}(F_1 F_2)$.

Corollary 1. *If $\mathscr{C}F_1$, $\mathscr{C}F_2$ and $F_1 F_2$ are connected, $\mathscr{C}(F_1 \cup F_2)$ is connected.* The extra condition "F_1 or F_2 is bounded" is again required in R^2, for 9·2 and Corollary 1, except when $F_1 F_2 = 0$. (*Example*: F_1 and F_2 are the positive ξ_1- and ξ_2-axes, including the origin in both cases.)

Corollary 2. *If D_1 and D_2 are domains in X^2 ($= Z^2$ or R^2) such that $D_1 \cup D_2 = X^2$, $D_1 D_2$ is connected.*

Theorem 9·3. *If a closed set F is contained in a domain D in X^2, and D_1, D_2, ... are the components of $\mathscr{C}F$, the components of $D - F$ are DD_1, DD_2, ..., provided $F \neq X^2$.*

DD_i *is not null.* For $\mathscr{C}F - D_i$ is open, since it is the union of components of $\mathscr{C}F$ (IV. 6·2), and therefore if DD_i were null

$$D_i \,|\, (\mathscr{C}F - D_i) \cup D$$

would be a partition of $\mathscr{C}F \cup D$, $= X^2$, which is impossible.

DD_i *is connected.* If x and y are any two of its points they are connected in D and in $\mathscr{C}F$, and $F \cap \mathscr{C}D = 0$. Therefore, by Theorem 9·2, x and y are connected in $D\mathscr{C}F$, i.e. they are joined by a continuum in D not meeting F, and therefore lying in D_i. Thus x and y are connected in DD_i.

Since the points of DD_i and DD_j, ($i \neq j$), are clearly not connected in $D - F$ the theorem is proved.

It follows that $\mathscr{C}F$ and $D - F$ have the same number of components.

Theorems 9·2 and 9·3 do not hold on surfaces in general. For example, on the ring-surface in Fig. 29 the circle γ determines one domain in the complete surface, but two in the white domain.

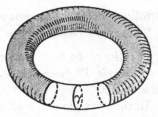

Fig. 29

Exercise. If the closed set F in X^2 has the components $F_1, F_2, ..., F_p$ (p finite), and if each F_i has a finite (positive) number, n_i, of residual domains, then $\mathscr{C}F$ has $\sum_{i=1}^{p} (n_i - 1) + 1$ components. [Induction on p.]

The following theorems form a partial converse of Alexander's Lemma.

Theorem 9·4·1. *Let G_1 and G_2 be open sets in X^2 such that $\mathscr{C}G_1$ and $\mathscr{C}G_2$ are connected. Let the 0-cycle Γ^0 in G_1G_2 bound the 1-chain κ_i in G_i, for $i = 1, 2$. Then if $\Gamma^0 \sim 0$ in G_1G_2, $\kappa_1 + \kappa_2 \sim 0$ in $G_1 \cup G_2$.*

Let $\Gamma^0 = \overset{\bullet}{\kappa}_0$, $|\kappa_0| \subseteq G_1 G_2$. Then the 1-cycle $\kappa_0 + \kappa_i$ does not meet the connected set $\mathscr{C}G_i$, and therefore bounds in G_i, and all the more in $G_1 \cup G_2$. Hence

$$(\kappa_1 + \kappa_2) = (\kappa_0 + \kappa_1) + (\kappa_0 + \kappa_2) \sim 0 \text{ in } G_1 \cup G_2.$$

Theorem 9·4·2. *Let F_1 and F_2 be connected closed sets in X^2 and let the 0-cycle Γ^0 in $\mathscr{C}(F_1 \cup F_2)$ bound the 1-chain κ_i not meeting F_i ($i = 1, 2$). Let the vertices of Γ^0 be paired in any way as*

$$(x_1, y_1), \ (x_2, y_2), \ ..., \ (x_q, y_q).$$

Then if $\kappa_1 + \kappa_2$ is non-bounding in $X^2 - F_1F_2$, at least one x_r is separated from y_r by $F_1 \cup F_2$.

If, for each vertex pair, $(x_r) + (y_r) \sim 0$ in $\mathscr{C}(F_1 \cup F_2)$, their sum $\Gamma^0 \sim 0$ in $\mathscr{C}(F_1 \cup F_2)$, and therefore $\kappa_1 + \kappa_2 \sim 0$ in $X^2 - F_1F_2$, by 9·4·1.

10. *A simple arc* (or *Jordan arc*) is a set of points homeomorphic with the closed segment $< 0, 1 >$. *A simple closed curve* (or *Jordan curve*) is a set of points homeomorphic with the circle $\xi_1^2 + \xi_2^2 = 1$ in R^2. The arc and curve are connected and compact, and are therefore closed sets in any containing space. It was shewn in IV. para. 2 that a simple arc or closed curve in R^2 or Z^2 is nowhere dense, and therefore if E is one or other of these curves,

$$\mathscr{F}E = \mathscr{F}(\mathscr{C}E) = E.$$

It is convenient in dealing with simple arcs and curves to suppose a definite topological mapping chosen in either case, and we may then speak of "the point τ of the curve", meaning in the case of the arc the image of the point τ of $< 0, 1 >$, and in the case of the closed curve the image of the point with angular coordinate

$2\pi\tau$ on the circle. In either case τ ranges from 0 to 1, but on the closed curve the points 0 and 1 are identical. The points 0 and 1 of the simple arc are called its *end-points*: it was proved on p. 62, Example 5, that they are distinguished topologically from other points of the curve.

Any point τ, other than 0 or 1, of the segment $< 0, 1 >$ divides it into two segments with τ as their only common point. It follows that if x is any point, other than an end-point, of a simple arc, the arc is the union of two simple arcs having no common point except x, which is an end-point of each. From the corresponding property of the circle it follows that if x and y are any two points of a simple closed curve, the curve is the union of two simple arcs having no common points except x and y, which are the end-points of both. It was shewn in IV. § 5 that these properties distinguish the simple arc and simple closed curve from all other compact connected spaces, and they are in fact the only properties that are used in proving the following theorems.

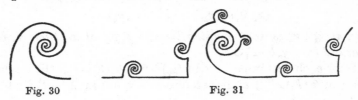

Fig. 30   Fig. 31

To understand the difficulties that would attend any attempt to deal with simple arcs and curves by means of linear approximations, and to appreciate the power of the "direct" method of Veblen, Brouwer and Alexander used in the following proofs, it should be borne in mind that these curves, although topologically "simple" in the sense that they have no multiple points, may from the point of view of differential geometry present highly condensed singularities. The arcs of two spirals $r = (\theta + \alpha_1)^{-\lambda}$ and $r = (\theta + \alpha_2)^{-\lambda}$ $(\lambda > 0, \theta > -\alpha_i)$, cut off by the circle $r = \beta$ form a simple arc with its end-points on the circle. A simple closed curve can be constructed having a spiral cusp of this kind in every arc, however small. The method of construction is the usual "condensation of singularities": first spiral cusps of a certain diameter ϵ_1 are introduced at regular intervals round a circle; then cusps of diameter ϵ_2 at closer intervals round this curve; and so on. It is easy to see that if the numbers ϵ_n tend to zero sufficiently fast the curves tend to a simple closed curve as limit.

Theorem 10·1. *A simple arc in the open or closed plane has a single complementary domain.*

Let L be the arc. Its complementary set is certainly not null, since L is nowhere dense (IV. 2).

Suppose the points x and y are separated by L. Then if a is the point $\tau = \frac{1}{2}$ of L, dividing it, say, into the two arcs L_1 and L_2, either L_1 or L_2 separates x from y. For if not, L_1 and L_2 satisfy the conditions of 9·2 ($L_1 L_2 = a$, and if the containing space is R^2 the arc, being compact, is bounded). Therefore their union, L, does not separate x and y.

A bisection argument now leads us to a contradiction. Of the two halves L_1, L_2 of L we select one that separates x and y, then a half of this half separating x and y, and so on, where a "half" of a simple arc means the image of half the pattern segment. If $L^{(n)}$ is the nth arc so chosen, the segments of which $L^{(1)}$, $L^{(2)}$, ... are images converge to a point of $< 0, 1 >$ (since each is a half of its predecessor). Therefore, owing to the continuity of the mapping, the arcs $L^{(1)}$, $L^{(2)}$, ... themselves converge to a point a_0 of L. For some n there is a square neighbourhood† of a_0 containing $L^{(n)}$ but not x or y. The points x and y are connected by a continuum not meeting this square, and therefore not meeting $L^{(n)}$.

Thus the assumption that x and y are separated by L has led to a contradiction.

Theorem 10·2. (Jordan's Theorem.)[17] *A simple closed curve in the open or closed plane has two com-* 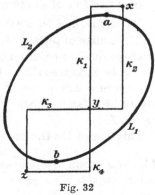 *plementary domains, of each of which it is the complete frontier.*

Let J be the curve. If the containing space is R^2, J is bounded.

I. *$\mathscr{C}J$ has at most two components.* Let a and b be any two points of J, dividing it into the simple arcs L_1 and L_2. Suppose that x, y and z are finite points of three different residual domains of J. By 10·1 there exist 1-chains κ_1 and κ_2 bounded by x and y, such that κ_i does not meet L_i

Fig. 32

† A square neighbourhood of the point at infinity is the *exterior* of any square.

($i = 1, 2$). Since x and y are separated by $L_1 \cup L_2$ it follows from Alexander's Lemma that $\kappa_1 + \kappa_2$ is non-bounding in $\mathscr{C}(L_1 L_2)$, i.e. in
$$X^2 - (a \cup b).$$

Similarly, $(y) + (z)$ is the common boundary of two 1-chains κ_3 and κ_4 not meeting L_1 and L_2 respectively, and $\kappa_3 + \kappa_4$ is non-bounding in $X^2 - (a \cup b)$. Therefore, by 6·4,

(1) $$(\kappa_1 + \kappa_2) + (\kappa_3 + \kappa_4) \sim 0 \quad \text{in} \quad X^2 - (a \cup b).$$

The 1-chain $\kappa_1 + \kappa_3$ does not meet L_1 and it is bounded by $(x + y) + (y + z)$, i.e. by x and z. Similarly, $\kappa_2 + \kappa_4$ does not meet L_2 and is bounded by x and z. The sum
$$(\kappa_1 + \kappa_3) + (\kappa_2 + \kappa_4)$$
is identical with (1), and therefore bounds in $\mathscr{C}(L_1 L_2)$. Hence x and z are not separated by $L_1 \cup L_2$, contrary to the hypothesis.

II. *$\mathscr{C}J$ has at least two components.* Let a, b, L_1, L_2 be as before and let Q be a square† with sides parallel to the coordinate axes separating a from b. Since the closed sets $L_1 Q$ and $L_2 Q$ do not meet, there is a grating **G** (which may be so chosen that Q is the locus of a 1-chain σ) no cell of which meets both $L_1 Q$ and $L_2 Q$. Let κ_0 be the set of 1-cells of σ that meet L_1.

The 0-cycle $\overset{\bullet}{\kappa}_0$ bounds κ_0 in $\mathscr{C}L_2$ and $\kappa_0 + \sigma$ in $\mathscr{C}L_1$, and
$$\kappa_0 + (\kappa_0 + \sigma) = \sigma$$
is non-bounding in $X^2 - (a \cup b) = X^2 - L_1 L_2$. Hence by 9·4·2 a pair of vertices of $\overset{\bullet}{\kappa}_0$ are separated by $L_1 \cup L_2$.

III. *If D_0 and D_1 are the components of $\mathscr{C}J$, $\mathscr{F}D_0 = \mathscr{F}D_1 = J$.* In the course of proving II it was shewn that there are points of both residual domains on Q. Now a is an arbitrary point of J and Q may be chosen so that it lies in an assigned neighbourhood of a. Therefore every point of J belongs to both $\mathscr{F}D_0$ and $\mathscr{F}D_1$. By IV. 3·1, both frontiers are contained in the frontier of $\mathscr{C}J$, i.e. in J.

This completes the proof of the whole theorem.

11. If J is bounded one of the residual domains contains all the points outside some circle. It is called the "outer domain", and denoted by D_o. The other, D_i, which is bounded, is called the

† I.e. the 1-dimensional frontier.

"inner domain". A set of points is said to be *inside*, or *outside*, J if it is contained in the inner, or outer, domain, respectively.

Theorem 11·1. *If J_1 is inside J_2, the inner domain of J_1 is inside J_2.*

Let D_{ri} and D_{ro} be the inner and outer domains of J_r. If D_{2o} met D_{1i} it would be contained in D_{1i}; for otherwise, being connected, it would meet $\mathscr{F} D_{1i}$, i.e. J_1, which is not so by hypothesis. But D_{2o} is unbounded, and therefore cannot be contained in D_{1i}. Hence D_{2o} does not meet D_{1i}, i.e. $D_{1i} \subseteq D_{2i}$.

Corollary. *If J_1 is inside J_2, J_2 is outside J_1,* for we have shewn that the inner domain of J_1 is inside J_2, and therefore does not meet J_2.

A domain whose frontier is a simple closed curve is called a *Jordan domain.*

The following are immediate consequences of Jordan's Theorem.

11·2. *If J is a simple closed curve in any domain D of X^2, $D - J$ has two components, whose frontiers are $J \cup F_1$ and $J \cup F_2$, where F_1 and F_2 are the parts of $\mathscr{F} D$ in the residual domains of J.* (Two components by 9·3, frontiers by II. 8, Example 2.) Hence

11·3. *Two disjoint simple closed curves have three complementary domains in X^2, of which two are Jordan domains, and the other has the two curves as frontier.*

By induction on the number of curves it follows that

11·4. *n disjoint simple closed curves in X^2 have $n + 1$ complementary domains.*

The following theorem is a direct generalisation of part II of Jordan's Theorem.

Theorem 11·5. *If the common part of two continua, F_1 and F_2, in Z^2 is not connected, there exists a pair of points not separated by F_1, but separated by $F_1 \cup F_2$.*

Let $H_1 \mid H_2$ be a partition of $F_1 F_2$, and on a grating, $\mathbf{G}$, of which no cell meets both H_1 and H_2 let K be the set of 2-cells that meet H_1. Since one of the 2-chains bounded by $\mathring{K}$ contains H_1 and the other H_2, $\mathring{K}$ is non-bounding in $Z^2 - F_1 F_2$.

Let $\mathbf{G}^*$ be a refinement of $\mathbf{G}$ such that no cell meets both $|\overset{*}{K}|F_1$ and $|\overset{*}{K}|F_2$ (which do not intersect), and let κ_0 be the set of edges of $\overset{*}{K}$, on $\mathbf{G}^*$, that meet F_2. Then $\overset{*}{\kappa}_0$ bounds κ_0 in $\mathscr{C}F_1$, and $\kappa_\bullet + \overset{*}{K}$ in $\mathscr{C}F_2$ (on $\mathbf{G}^*$); and $\kappa_0 + (\kappa_0 + \overset{*}{K}) = \overset{*}{K}$ is non-bounding in $Z^2 - F_1F_2$. Since each component of κ_0 contains an even number of boundary vertices, the vertices of $\overset{*}{\kappa}_0$ can be named $x_1,\ y_1,\ x_2,\ y_2,\ \ldots,\ x_q,\ y_q$ so that x_r and y_r are for each r connected in κ_0 and therefore in $\mathscr{C}F_1$. But by 9·4·2 at least one pair $x_r,\ y_r$ are separated by $F_1 \cup F_2$.

(In R^2 a boundedness condition must be added, e.g. that F_1F_2 be bounded.)

Corollary. *If the union of the two continua, F_1 and F_2, is Z^2, F_1F_2 is connected (Z^2 is "unicoherent").*

If D is a domain in X^2, a simple arc, L, with one end-point on $\mathscr{F}D$ and all its other points in D, is called an *end-cut*. If both end-points are on $\mathscr{F}D$ and the rest in D the arc is a *cross-cut*.

Theorem 11·6. *If L is an end-cut in D, $D - L$ is connected and its frontier is $\mathscr{F}D \cup L$.*

That $D - L$ is connected follows from 9·2, applied to L and $\mathscr{C}D$. Since $D - L = D \cap \mathscr{C}L$ its frontier is contained in $\mathscr{F}D \cup \mathscr{F}(\mathscr{C}L)$, $= \mathscr{F}D \cup L$. Any small enough neighbourhood of a point b, of $\mathscr{F}D$ other than a (the end-point) contains points of D but none of L, and therefore b belongs to $\mathscr{F}(D - L)$. Similarly, every point of L except a is in $\mathscr{F}(D - L)$. Finally a, being a point of closure of $L - (a)$, belongs to $\mathscr{F}(D - L)$.

Theorem 11·7. *If both the end-points of a cross-cut L in a domain D of X^2 are on the same component of $\mathscr{C}D$, $D - L$ has two components, and L is contained in the frontiers of both.*

The proof that $D - L$ has at most two components is exactly like part I of Jordan's Theorem, the closed sets L and $\mathscr{C}D$ (which each determine only one domain) taking the place of L_1 and L_2.

To prove the rest suppose first that $\mathscr{C}D$ is connected. Then by 11·5 $D - L$, $(= \mathscr{C}(L \cup \mathscr{C}D))$, has at least two components, and it follows, just as in part III of Jordan's Theorem, that the end-points, a and b, of L belong to the frontiers of both. If c is any other point of L (Fig. 33), the set (arc $ac \cup \mathscr{C}D$) determines only one domain, in which arc bc is a cross-cut. Therefore, by the case just settled, c belongs to both frontiers.

In a general domain, if F is the component of $\mathscr{C}D$ meeting L, $\mathscr{C}F$ is connected, by IV. 3·3, and by the previous case $\mathscr{C}F - L$, $= \mathscr{C}(F \cup L)$, has two components, both having L in their frontiers. If x is a point of L, other than an end-point, the points of $\mathscr{C}L$ in a sufficiently small neighbourhood of x all belong to $D - L$, and include pairs of points which are separated by $F \cup L$, a fortiori by $\mathscr{C}D \cup L$. Hence $D - L$ has two components and L is in the frontier of both.

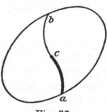 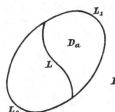

Fig. 33   Fig. 34

Theorem 11·8. *If D_1 is one of the complementary domains in X^2 of a Jordan curve, J, and L a cross-cut in D_1 whose end-points divide J into the arcs L_1 and L_2, the components of $D_1 - L$ have the frontiers $L \cup L_1$ and $L \cup L_2$.*

Since $L \cup L_1$ does not meet the other complementary domain, D_0, of J, one of the residual domains of $L \cup L_1$, say D_a, does not meet D_0 (Fig. 34). Thus $D_a \subseteq X^2 - D_0$, and since D_a is open,

$$D_a \subseteq \mathscr{I}(X^2 - D_0) = D_1.$$

Hence D_a, which does not meet L, is contained in $D_1 - L$, and therefore in one of its components, say D_α. If D_α contained a point of $\mathscr{C}D_a$ it would meet $\mathscr{F}D_a$, i.e. $L \cup L_1$, which is impossible, since $D_\alpha \subseteq \mathscr{C}(L \cup L_1 \cup L_2)$. Thus $D_\alpha = D_a$ and $\mathscr{F}D_\alpha = L_1 \cup L$. Similarly, $L_2 \cup L$ is the frontier of a component of $D_1 - L$. Since there are only two components this proves the theorem.

Exercises. 1. If D_1 and D_2 are domains in Z^2 and $\mathscr{F}D_1 \cap \mathscr{F}D_2$ is connected $D_1 D_2$ is connected.

2. If D_1 and D_2 are domains in Z^2 and $\mathscr{F}D_1 \cap \mathscr{F}D_2$ has not more than two components, $D_1 D_2$ has not more than two components. [Follow part I of Jordan's Theorem.]

3. The frontier of a domain D is a pair of non-intersecting Jordan curves, J_1 and J_2, and L_1 and L_2 are non-intersecting cross-cuts in D, each having one end-point on J_1 and one on J_2. Shew that $D - (L_1 \cup L_2)$ has two components, each of which is a Jordan domain. [Let the end-points of L_1 and L_2 on J_1 divide it into the arcs L_3 and L_4. Consider first $L_1 \cup L_3 \cup L_2$ as a cross-cut in one of the J_2-domains (say D_0), and then L_4 as a cross-cut in a component of $D_0 - (L_1 \cup L_3 \cup L_2)$.]

4. A cross-cut in a domain D in Z^2 with its end-points in different components of $\mathscr{F}D$ does not destroy the connection of D. [If L is the cross-cut and $H_1 \mid H_2$ a partition of $\mathscr{F}D$ separating the end-points of L choose two points of $D - L$ and apply 9·2 first to H_1 and L, then to $H_1 \cup L$ and H_2. See 16·3 for another proof. This theorem fails in R^2 unless $\mathscr{F}D$ is bounded.]

5. If F_1, F_2, ..., F_p are closed sets in Z^2, such that, for every i, j and k, $\mathscr{C}F_i$ and $F_i F_j$ are connected, and $F_i \cup F_j \cup F_k \neq Z^2$, then $F_1 F_2 ... F_p$ is connected. [It is sufficient to prove the case $p = 3$, for $F_1 F_p, F_2 F_p, ..., F_{p-1} F_p$ are then seen to be $p - 1$ sets fulfilling the conditions, and the theorem follows by induction. Suppose $p = 3$, and that the theorem is false. Then $F_1(F_2 \cup F_3)$ determines at least two domains, and it is easily shewn that none of these domains can be contained in $F_2 \cup F_3$. Therefore there are two points not in $F_2 \cup F_3$ and separated by it. This contradicts 9·2.

6. Prove the same result when "closed" is replaced by "open". [Use 11·5 and 9·2.][18]

§ 3. INVARIANCE OF DIMENSION NUMBER AND OF OPEN SETS

12. Theorem 12·1. *If $p \neq 2$, R^p cannot be mapped topologically on to R^2, nor Z^p on to Z^2.*

The value $p = 0$ is obviously impossible, and $p = 1$ has already been disposed of (ɪᴠ. para. 2). Suppose then that $p > 2$.

Let C be the unit circle in the (ξ_1, ξ_2)-plane in R^p. C does not destroy the connection of R^p, for any point x of $R^p - C$ can be joined to o in $R^p - C$ by at most two segments in the plane through x and the ξ_3-axis. Thus if f is a function that maps R^p topologically on to R^2, $R^p - C$ is connected, but its image $R^2 - f(C)$ is the complement in R^2 of a simple closed curve, and therefore not connected. This contradicts the proved topological invariance of connectedness.

If Z^p and Z^2 were homeomorphic they could be correlated so that the points at infinity corresponded. This would give a homeomorphism between R^p and R^2, which has just been proved impossible.

A 2-*element* is any homeomorph of a bounded 2-cell, i.e. of a closed rectangular region in the plane.

Theorem 12·2. *The complement of a 2-element in X^2 ($= Z^2$ or R^2) is connected and not null.*

Let F, the 2-element, be the image of the 2-cell A (with frontier Q) under the topological mapping f. F is not the whole of X^2, for $A - Q$ is connected, but $f(Q)$, a simple closed curve, destroys the connection of X^2.

Let A be divided into two equal 2-cells, A_1 and A_2, by a segment α parallel to one side. If two points x and y of $X^2 - F$ are connected both in $X^2 - f(A_1)$

Fig. 35

and in $X^2 - f(A_2)$ they are connected in $X^2 - F$, by 9·2.

The rest of the proof follows that of 10·1. If x and y are separated by F let the cell A be repeatedly bisected, by segments alternately parallel to the ξ_1- and ξ_2-axes. Then there exists a sequence of 2-cells,

$$A_0 (= A), A_1, A_2, \ldots,$$

each a half of the one before, such that $f(A_n)$ separates x and y. But since the diameter of $f(A_n)$ tends to zero this is impossible (cf. p. 115), and therefore no pair of points separated by F exists.

13. The theorem of the *invariance of open sets* depends on the following lemma:

Theorem 13·1. *If A is a bounded 2-cell on a grating, and f a $(1, 1)$ continuous mapping of A into X^2, $f(\mathscr{I}A)$ is one of the residual domains of the simple closed curve $f(Q)$, $(Q = \mathscr{F}A)$.*

The $(1, 1)$ continuous mapping f of the compact sets Q and A is, by III. 9·3, a homeomorphism. Therefore $f(Q)$ is a simple closed curve and $f(A)$ a 2-element. $f(\mathscr{I}A)$ is a connected set which does not meet the simple closed curve $f(Q)$ and therefore lies in one of its residual domains, say D_1. If it is not the whole of D_1

there are points of $X^2 - f(A)$ in both the residual domains of $f(Q)$, but not on the frontier $f(Q)$ itself. This is impossible, since $X^2 - f(A)$ has been shewn in 12·2 to be connected.

Theorem 13·2. (Invariance of open sets.) *If an open set G in X^2 is mapped $(1, 1)$ continuously on to the set E in X^2, E is also an open set in X^2.*

First suppose that G is not the whole of the closed plane Z^2. Then it can be arranged, by a topological mapping of the whole plane on to itself, that G does not contain the point at infinity. If then y is any point of E, say $y = f(x)$, there is a closed bounded 2-cell contained in G and containing x. By the preceding lemma the interior of this cell is mapped on to an open set, which is contained in E and contains y. Hence E is an open set.

If G is the whole of Z^2, every neighbourhood except the whole space is mapped on to an open subset of E, by the case just settled. Therefore E is again open.

This theorem is not to be confused with the proposition that if X^2 is mapped topologically on to itself open sets are mapped on to open sets, which follows immediately from the definition of a topological mapping. In the theorem just proved it is not assumed that the mapping is defined outside G, nor that there exists a topological mapping of the whole space on to itself coinciding with the given one in G.

Corollary 1. *A $(1, 1)$ continuous mapping f of an open set in X^2 on to any set E in X^2 is a homeomorphism.* For the f-image of any open set is open in X^2, a fortiori in E, and therefore f^{-1} is continuous (III. 9·2).

Corollary 2. *If f maps any set E_1 of X^2 topologically on to E_2 of X^2, it maps $\mathscr{I}(E_1)$ on to $\mathscr{I}(E_2)$.* If x is a point of $\mathscr{I}E_1$ there is an open set G containing x and contained in E_1, and $f(G)$ is an open set contained in E_2 and containing $f(x)$. Hence $f(\mathscr{I}E_1) \subseteq \mathscr{I}E_2$, and similarly $f^{-1}(\mathscr{I}E_2) \subseteq \mathscr{I}E_1$.

Corollary 3. *If in Corollary 2 E_1 and E_2 are closed, $\mathscr{F}E_1$ is mapped on to $\mathscr{F}E_2$.* For $\mathscr{F}E_i = E_i - \mathscr{I}E_i$.

Corollary 4. *Z^2 is not the homeomorph of any proper subset of itself.* For if the subset E is homeomorphic with the whole space,

E is open, by 13·2, and closed, because as the topological image of a compact space it is compact. Therefore, since Z^2 is connected, $E = Z^2$.

Exercise. Prove that if $p > 2$ no non-null open set G in R^p can be mapped topologically on to a set of points in R^2. [We may suppose G contains the origin. If E_{12} is the (ξ_1, ξ_2)-plane in R^p, GE_{12} is, by 13·2, mapped on to an open subset of R^2. But the points of GE_{12} are limit points of $G - GE_{12}$.]

§ 4. FURTHER SEPARATION THEOREMS

14. A number of important results are derivable from the two following simple theorems. It is to be observed that, although these theorems hold in R^2 as well as in Z^2, no compactness condition is imposed.

Theorem 14·1. *If the domains D_1 and D_2 in the open or closed plane X^2 do not meet, but $\mathscr{F}D_1 \subseteq \mathscr{F}D_2$, then $\mathscr{F}D_1$ is connected.*

No closed proper subset, F, of $\mathscr{F}D_1$ separates D_1 from D_2, for if x is a point of $\mathscr{F}D_1 - F$ there are points y_1 and y_2 of D_1 and D_2 nearer to x than to F, and the segment $y_1 y_2$ does not meet F. If, then, $H_a \mid H_b$ is a partition of $\mathscr{F}D_1$, neither H_a nor H_b separates D_1 from D_2, and therefore, since they do not meet, their union, $\mathscr{F}D_1$, does not separate D_1 and D_2 (9·2; since $H_a \cap H_b = 0$ no compactness condition is required). This is contrary to the hypothesis that D_1 and D_2 do not meet.

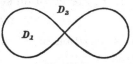

Fig. 36

Theorem 14·2. *If D is any domain in X^2 the components of the open set $\mathscr{C}\bar{D}$ all have connected frontiers.*

If D_0 is a component of $\mathscr{C}\bar{D}$

$$\mathscr{F}D_0 \subseteq \mathscr{F}(\mathscr{C}\bar{D}) \qquad \text{(IV. 3·1)}$$
$$= \mathscr{F}(\bar{D}) \subseteq \mathscr{F}D:$$

the conditions of 14·1 are satisfied.

Theorem 14·3. *If x and y are separated by the closed set F in the open or closed plane they are separated by a component of F.*

If D_x is the component of $\mathscr{C}F$ containing x, y belongs to a component, D_y, of $\mathscr{C}\bar{D}_x$, and $\mathscr{F}D_y$ is, by 14·2, a connected subset of F

separating x and y. Hence the component of F containing $\mathscr{F}D_y$ separates x and y.

Theorem 14·4. *Every residual domain of a connected closed set in the open or closed plane has a connected frontier.*

Let F be the connected closed set, D a residual domain, and suppose that $H_1 \mid H_2$ is a partition of $\mathscr{F}D$.

Consider the open set $\mathscr{C}\bar{D}$. Since $\mathscr{F}(\mathscr{C}\bar{D}) = \mathscr{F}\bar{D} \subseteq \mathscr{F}D$ there is no component of $\mathscr{C}\bar{D}$ whose frontier meets both H_1 and H_2, for all these frontiers are connected, by 14·2. Thus the components of $\mathscr{C}\bar{D}$ can be sorted into two classes, with frontiers in H_1 and H_2 respectively. If the unions of these two classes of open sets are G_1 and G_2, $G_1 \cup H_1$ *is a*

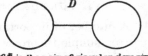

$\mathscr{C}\bar{D}$ *is the pair of circular domains*

Fig. 37

closed set. For every neighbourhood of a point x of $\mathscr{F}G_1$ contains a point of some component, D_1, of the G_1-class, and therefore also (since x is not itself in D_1) a point of $\mathscr{F}D_1$, i.e. of H_1. Therefore $\mathscr{F}G_1 \subseteq \bar{H}_1 = H_1$, and $G_1 \cup H_1 = \bar{G}_1 \cup H_1$, a closed set.

Thus $G_1 \cup H_1$ and $G_2 \cup H_2$ are disjoint closed sets, whose union is $\mathscr{C}\bar{D} \cup \mathscr{F}D = \mathscr{I}\mathscr{C}D \cup \mathscr{F}\mathscr{C}D = \mathscr{C}D$, and contains F. Since H_1 and H_2 are non-null subsets of F,

$$(G_1 \cup H_1)F \mid (G_2 \cup H_2)F$$

is a partition of F, contrary to the assumption that F is connected.

Corollary. *If D is any domain in the open or closed plane the residual domains of $\mathscr{F}D$, other than D itself, have connected frontiers.* For these domains are precisely the residual domains of the connected set $\bar{D}$, since

$$\mathscr{C}(\mathscr{F}D) = \mathscr{C}(\bar{D} - D) = \mathscr{C}\bar{D} \cup D.$$

Theorem 14·5. *If D is a domain in the open or closed plane, each component of $\mathscr{C}D$ contains just one component of $\mathscr{F}D$.*

Since $\mathscr{F}D \subseteq \mathscr{C}D$ every component of $\mathscr{F}D$ is contained in a component of $\mathscr{C}D$. It is therefore sufficient to prove that the part of $\mathscr{F}D$ in any component, C, of $\mathscr{C}D$ is connected and not null.

The part of $\mathscr{F}D$ in C is in fact $\mathscr{F}C$ (IV. 6·4), and therefore certainly not null. By Theorem 3·3 of Chapter IV, $\mathscr{C}C$ is a domain, and therefore by 14·4 $\mathscr{F}C$ is connected.

From this theorem it follows that *if a connected set meets two components of $\mathscr{F}D$ it meets D*, for otherwise the two components would be connected in $\mathscr{C}D$.

By a special case of 14·5 the properties of having a connected *frontier*, and of having a connected *complement*, are equivalent for domains in R^2 or Z^2. In the closed plane domains with these properties are identical with the *simply connected* domains that are considered in the next chapter. Simpler proofs of the equivalence of the two properties, and of 14·4, valid only in the closed space Z^2, will be given in Chapter VI (para. 4).

Exercise. Prove that if D is a domain in Z^2 a cut-point of a component of $\mathscr{F}D$ is a cut-point of a component of $\mathscr{C}D$. Is this true of domains on a ring-surface? [Take the cut-point to be the point at infinity.]

15. Although the theorems that have been proved in para. 14 resemble certain theorems established in Chapter IV, para. 5, they are sharply distinguished from them by the fact that the new theorems hold only in a restricted class of spaces, which includes R^2 and Z^2, but excludes such a simple surface as the torus or ring-surface. If F is a cylindrical region on a torus, the complement of F is a domain, D, whose frontier consists of two disjoint circles, both lying in the same component of $\mathscr{C}D$. Thus Theorems 14·4 and 14·5 fail for the torus. Any one circle running round the ring leaves it connected, but two such circles destroy the connection: Theorem 14·3 fails.

16. The following generalisation of Alexander's Lemma will often be found effective in solving separation problems that cannot easily be settled by the original Lemma.

Theorem 16·1. *With the notation of Alexander's Lemma in Z^2, 9·1·2, if $\kappa_1 + \kappa_2$ bounds in $Z^2 - F_1P$, for each component, P, of F_2, x and y are not separated by $F_1 \cup F_2$.*

That this is a stronger theorem than the original Lemma is shewn by the example illustrated in Fig. 38. If F_1 is the cross-cut from a to b, and F_2 the complement of the annular white domain, F_1F_2 (i.e. the point-pair a, b) prevents $\kappa_1 + \kappa_2$ from bounding, and nothing follows from the Lemma. But the intersection of F_1 with each component of F_2 is a single point, which cannot prevent any cycle from bounding. Therefore from the new theorem it follows that $(x) \sim (y)$ in $\mathscr{C}(F_1 \cup F_2)$.

Proof. Let the two 2-chains bounded by $\kappa_1 + \kappa_2$ be called the "white" chain (K_w) and the "black" chain (K_b). It is given that the intersection of F_1 with any component of F_2 lies either entirely inside the black chain or entirely inside the white, and we may classify components of F_2 that meet F_1 into "black" and "white" components accordingly.

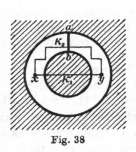

Fig. 38

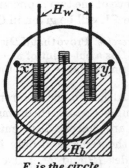

F_1 is the circle,
F_2 the three thick vertical lines

Fig. 39

The union, X, of the black components is closed. Let a be any point of $\bar{X}$, and $x_1, x_2, \ldots$ a sequence of points of X converging to a, say $x_n \in X_n$, where X_n is a black component. By the definition of these components X_n contains a point y_n of the compact set $F_1 | K_b |$, and a subsequence (y_{n_r}) converges to some point b of $F_1 | K_b |$. By IV. 5·7 the points a and b are connected in F_2. But the component containing b is certainly "black", since b itself is in K_b. Therefore a is in a black component, i.e. $\bar{X} \subseteq X$.

Similarly, the union Y of the white components is closed, and therefore, since no component of F_2 can meet both X and Y, there exists a partition $H_b | H_w$ of F_2 such that $X \subseteq H_b$ and $Y \subseteq H_w$ (IV. 5·6).

The method used in proving the Alexander Lemma is now applied to H_b and H_w separately. A rectangular grating, **G**, is chosen such that no cell meets (1) both F_1 and $H_b \,|\, K_w|$, nor (2) both F_1 and $H_w \,|\, K_b\,|$, nor (3) both H_b and H_w. Let K_1 be the sum of the 2-cells of K_w that meet H_b, and K_2 the sum of the 2-cells of K_b that meet H_w. Then $K_w + K_1 + K_2$ contains no cell meeting H_b, and all cells of $\mathbf{G}$ meeting H_w; and it follows that its boundary does not meet F_2. Hence, as in the original Lemma, if

$$\gamma = \overset{\bullet}{K}_1 + \overset{\bullet}{K}_2$$

$\kappa_1 + \gamma$ is a 1-chain, bounded by x and y, not meeting F_1 or F_2. The theorem is therefore proved.

In R^2 the condition "F_2 is bounded" must be added.

Theorem 16·2. *If F_1 and F_2 are closed sets in Z^2, and the intersection of F_1 with each component of F_2 is connected, two points which are connected in $\mathscr{C}F_1$ and $\mathscr{C}F_2$ are connected in $\mathscr{C}(F_1 \cup F_2)$.* Follows immediately from 16·1.

Corollary. *If the intersection of a closed set F in Z^2 with each component of $\mathscr{F}D$ is connected, two points of D which are connected in $\mathscr{C}F$ are connected in $D - F$.* By 16·2 the points are connected by a continuum which does not meet $\mathscr{F}D$ or F, and therefore lies in $D - F$.

Theorem 16·3. *A cross-cut in a domain D of Z^2 with its end-points on different components of $\mathscr{F}D$ does not destroy the connection of D.*

In the open plane the condition "F_2 is bounded" must be added in 16·2, and "$\mathscr{F}D$ is bounded" in the Corollary and 16·3.

Other proofs of these deductions from 16·1 will be given in Chapter VI.

It would be natural to expect the following further generalisation of the Alexander Lemma to hold: *with the notation of the original Lemma, if for each component P_1 of F_1 and P_2 of F_2 the set $P_1 P_2$ does not prevent $\kappa_1 + \kappa_2$ from bounding, then x and y are connected in $\mathscr{C}(F_1 \cup F_2)$.* That this theorem is false is shewn by the simple example illustrated in Fig. 40. The horizontal lines are the components of F_1, the verticals those

Fig. 40

of F_2; the intersection of any two components is a point, which cannot prevent $\kappa_1 + \kappa_2$ from bounding, but x is separated from y by $F_1 \cup F_2$.

Exercises. 1. If F is the union of an infinity of simple arcs in X^2, with a common end-point but otherwise non-intersecting, $\mathscr{C}F$ is connected. [If x and y are separated by F, cut off the part of F near the common end-point, a, by a point near a in each of the arcs, and apply 16·2.]

2. Let x and y be points of Z^2, F_1 a continuum, and F_2 any closed set such that no set $F_1 \cup C$ separates x and y, if C is a component of F_2. Then $F_1 \cup F_2$ does not separate x and y. [Neither F_1 nor (by 14·3) F_2 alone separates x and y. Let $(x)+(y)$ bound κ_i in $\mathscr{C}F_i$. By 9·4·1 $\kappa_1 + \kappa_2 \sim 0$ in $Z^2 - F_1 C$.]

3. Deduce 16·2 from Exercise 2. [Since each component of F_2 meets at most one of F_1, each component of $F_1 \cup F_2$ is either a component of F_2, or the union of a component of F_1 and the components of F_2 that meet it. Use Example 1, p. 83.]

§5. EXTENSION TO SETS OF POINTS IN Z^p AND R^p

The arguments of §§ 1–4 (unlike those that will be used in Chapters VI and VII) are not essentially 2-dimensional in character, and it will now be shewn how a large part of the theory of these sections may be extended to p-dimensional space. The definitions and properties of gratings and cells require more formal treatment, since they can no longer be seen by the inspection of two or three possible cases in a diagram, but the rest of the theory is little altered.

The results of this section are not used in the rest of the book.

17. The inductive proof of the "fundamental lemma" (5·1) started in two dimensions from a grating consisting of a single square, but in p dimensions the analogous "cube" is already too complicated a figure to form a convenient starting-point. We therefore use a grating consisting of *complete* linear $(p-1)$-spaces, and not merely the parts cut off in a "cube". A typical grating of the new kind in two dimensions is illustrated in Fig. 41.

We first define gratings in the open space R^p.

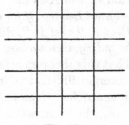

Fig. 41

The name *principal* is used for a set of points determined by one equation $\xi_r = \alpha$. A finite set of principals in R^p is a *rectangular grating*, **G**, if it contains at least one principal $[\xi_r = \alpha_r]$ for each $r = 1, 2, 3, ..., p$. We suppose the principals numbered in any way as $\pi_1, \pi_2, ..., \pi_m$.

Any principal $\xi_r = \alpha$ determines three closed convex sets, namely the principal itself, and the two *closed half-spaces* $\xi_r \leqslant \alpha$ and $\xi_r \geqslant \alpha$. If, for each principal π_i of **G**, one of these three sets is chosen and called P_i, the common part

$$P_1 P_2 \ldots P_m,$$

if not null, is called a *cell*; and it is a *k-cell* if just $p - k$ of the P_i are principals. Since the P_i are closed convex sets all cells are closed and convex; and since every point of R^p belongs, for each i, to at least one closed half-space determined by π_i, the union of the p-cells is R^p.

It will be noticed that in a product $P_1 P_2 \ldots P_m$ there are in general superfluous factors. For example, if P_1 is $\xi_1 = 1$ and P_2 is $\xi_1 \leqslant 2$, then whatever $P_3, ..., P_m$ may be,

$$P_1 P_2 P_3 \ldots P_m = P_1 P_3 \ldots P_m.$$

The formal treatment is, however, greatly simplified by retaining these superfluous factors, so that in the product defining any cell there appears just one factor corresponding to each principal of the grating.

A cell is determined by m equations and inequalities, each restricting only one coordinate, and is therefore the set formed by letting each coordinate range independently through a certain set of values. If the defining relations include $\xi_r \leqslant \alpha$ they do not include $\xi_r = \alpha$ or $\xi_r \geqslant \alpha$, since there is only one "factor" for each principal of **G**. The range of values of ξ_r cannot therefore in this case reduce to α, and hence includes at least one value in $\xi_r < \alpha$. In particular, in a p-cell, where all the factors of the defining "product"† are closed half-spaces, the "product" of the corresponding *open* half-spaces—an open set contained in the cell—is not null, and the p-cell itself is the closure of this open set.

It follows that, for $k > 0$, a k-cell is the closure of an open set in some $[k]$, and that a 0-cell is a point.

† The intersection or $\cap$-product.

The following are simple consequences of the definitions.

(A) The only cells containing a cell $P_1 P_2 \ldots P_m$ are those obtained by replacing certain P_i which are principals by corresponding closed half-spaces. For in any other change at least one half-space $\xi_r \leqslant \alpha$ is replaced by $\xi_r \geqslant \alpha$ or $\xi_r = \alpha$. In either case the set no longer contains a point of $\xi_r < \alpha$ and therefore by the last paragraph, does not contain the given cell.

In particular, the specification of a cell in the form $P_1 P_2 \ldots P_m$ is unique.

(B) The only cells contained in $P_1 P_2 \ldots P_m$ are those obtained by replacing half-spaces by the corresponding principals.

(C) *If the k-cell A^k is contained in the principal π there are just two $(k+1)$-cells that contain A^k but are not contained in π.* The cell A^k is of the form πX, where X is a certain "product". The only $(k+1)$-cells containing A^k are those whose specification is obtained by changing just one factor of πX from a principal to a half-space. If the factor π is not changed the resulting set is in π. Therefore the only cells of the required type are $P_+ X$ and $P_- X$, where P_+ and P_- are the closed half-spaces determined by π. These cells are different from each other, since, as we have seen, they contain points on different sides of π.

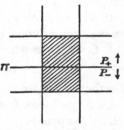

$(p=2,\ k=1)$ X *is shaded*

Fig. 42

In particular, every $(p-1)$-cell is contained in just two p-cells.

Gratings in Z^p. A rectangular grating in Z^p is, by definition, a rectangular grating in the subspace R^p. The k-cells are for $k > 0$ the closures in Z^p of the k-cells in R^p. The 0-cells are the finite 0-cells of the R^p-grating and w, the point at infinity. The symbol $P_1 P_2 \ldots P_m$ may still be used to denote any cell except w, and on this understanding (A), (B) and (C) remain true, except that w belongs to all infinite† cells and no finite ones.

We now confine ourselves to gratings in Z^p.

18. A *k-chain* on a grating **G** is any set of k-cells of **G**. The *sum* $C_1^k + C_2^k$ of the chains C_1^k and C_2^k is the set of k-cells belonging

† See footnote, p 101.

to one but not both of them. This addition is associative and commutative. Ω^p denotes the set of all p-cells, and

$$\mathscr{C}C^p = \Omega^p + C^p.$$

The *boundary*, $\mathring{A}^k$, of a k-cell A^k ($1 \leqslant k \leqslant p$) is the set of $(k-1)$-cells contained in it, and $\mathring{C}^k$ is the sum of the boundaries of the k-cells of C^k. It follows immediately that $(C_1^k + C_2^k)^\bullet = \mathring{C}_1^k + \mathring{C}_2^k$. A k-*cycle* is, if $k > 0$, a k-chain with zero boundary, and, if $k = 0$, an even number of vertices (0-cells). The sum of any set of k-cycles is evidently a k-cycle.

Theorem 18·1. *If* $k > 0$, $\mathring{C}^k$ *is a* $(k-1)$-*cycle*.

It is evidently sufficient to prove the theorem when C^k contains only one cell.

(i) Let $k = 1$. From the condition that the grating contains at least one principal parallel to $[\xi_r = 0]$, for each r, it follows that every 1-cell contains two vertices (one of which may be w).†

(ii) Suppose $k > 1$. It follows from remark (B) that if

$$A^{k-2} \subseteq A^k,$$

then $A^{k-2} = \pi_1 \pi_2 X$ and $A^k = Q_1 Q_2 X$, where π_i and Q_i are corresponding principal and half-space. The only $(k-1)$-cells containing $\pi_1 \pi_2 X$ and contained in $Q_1 Q_2 X$ are $\pi_1 Q_2 X$ and $\pi_2 Q_1 X$. Thus every $(k-2)$-cell of $\mathring{A}^k$ is contained in just two $(k-1)$-cells of $\mathring{A}^k$, which is therefore a cycle.

The *locus* $|C^k|$ of C^k is the union of its k-cells. C^k is *connected* if $|C^k|$ is connected, and the components of any chain C^k are the subchains whose loci are the components of $\lfloor C^k \rfloor$. Theorems 3·1 and 3·2, and the analogue of 3·3 ($\mathscr{F} \lfloor C^p \rfloor = \mathring{C}^p$) remain true, and are similarly proved.

A grating $\mathbf{G}^*$ in Z^p is a *refinement* of a grating $\mathbf{G}$ if every principal of $\mathbf{G}$ is a principal of $\mathbf{G}^*$. Clearly any two gratings have a common refinement. Suppose that $\pi_1, \pi_2, \ldots, \pi_m$ are the principals of $\mathbf{G}$, and that $\mathbf{G}^*$ is formed by adding a single new principal π_{m+1}, with half-spaces Q_{m+1}^+ and Q_{m+1}^-. The cells of $\mathbf{G}^*$ (other than

---

† This case of 18·1 fails for chains on a grating in R^p, and with it the important theorem that if $\mathring{C}^1 = (x) + (y)$, x and y are connected in the locus of C^1.

w) are the non-null sets whose product symbols are obtained by appending an extra factor π_{m+1}, Q^+_{m+1} or Q^-_{m+1} to the symbol of a cell of $\mathbf{G}$, and may accordingly be denoted by $A^k\pi_{m+1}$, $A^kQ^+_{m+1}$ or $A^kQ^-_{m+1}$. Two of these products may be empty, and the third is then A^k itself. Thus each A^k of $\mathbf{G}$ is represented on $\mathbf{G^*}$ by a k-chain A^k_* consisting either of A^k itself or of the two k-cells $A^kQ^+_{m+1} + A^kQ^-_{m+1}$. We may write symbolically (in either case) $A^k_* = A^k(Q^+_{m+1} + Q^-_{m+1})$. It follows that each k-chain C^k on $\mathbf{G}$ is represented by a *subdivided chain* C^k_*, the sum of the subdivided cells of C^k, and therefore given symbolically by $C^k(Q^+_{m+1} + Q^-_{m+1})$. Clearly $|C^k_*| = |C^k|$, and for any two chains,

$$(C^k_1 + C^k_2)_* = C^k_{1*} + C^k_{2*}.$$

If $k > 0$ a $(k-1)$-cell of $A^kQ^+_{m+1}$ is, by (B), to be found by substituting a principal for a half-space factor, which may be either a factor of A^k or Q^+_{m+1} itself.

Hence

$$(A^kQ^+_{m+1})^\bullet = \mathring{A}^kQ^+_{m+1} + A^k\pi_{m+1},$$

and similarly,

$$(A^kQ^-_{m+1})^\bullet = \mathring{A}^kQ^-_{m+1} + A^k\pi_{m+1}.$$

Hence, adding

$$(A^k_*)^\bullet = (A^kQ^+_{m+1})^\bullet + (A^kQ^-_{m+1})^\bullet$$
$$= \mathring{A}^k(Q^+_{m+1} + Q^-_{m+1}) = (\mathring{A}^k)_*.$$

By summation we find $(C^k_*)^\bullet = (\mathring{C}^k)_*$.

The results thus established for a refinement by adjunction of a single principal may evidently be extended step by step to any refinement, and we have therefore proved

(D) If $\mathbf{G^*}$ is a refinement of $\mathbf{G}$, there corresponds to each k-chain C^k on $\mathbf{G}$ a *subdivided chain* C^k_*, such that $|C^k_*| = |C^k|$, $(C^k_1 + C^k_2)^* = C^k_{1*} + C^k_{2*}$, and (if $k > 0$) $(C^k_*)^\bullet = (\mathring{C}^k)_*$.

This justifies the introduction of the convention of para. 6, by which the subdivided chain C^k_* on $\mathbf{G^*}$ is denoted by the same symbol C^k as its original on $\mathbf{G}$.

The definitions of "$\Gamma^k \sim 0$ in G" and "$\Gamma^k_1 \sim \Gamma^k_2$ in G" are the same as for the 2-dimensional case and Theorem 6·1 and its corollary still hold, and are similarly proved.

19. The following slightly stronger form of the 2-dimensional "Fundamental Lemma" is required (the only substantially new argument in this section).

Theorem 19·1. *Let Γ^k be a k-cycle on a grating in Z^p, and let a be a point not on Γ^k. If $0 \leqslant k \leqslant p-1$, $\Gamma^k \sim 0$ in $Z^p - (a)$.*

We may suppose $k > 0$, since the result is obvious if $k = 0$. It may also be assumed (by a preliminary refinement of the grating, G_1, if necessary) that a is a vertex. The proof is by induction on the number of principals of G_1, and is trivial if G_1 is a grating G_σ of the simplest type, with one principal in each allowed direction; for G_σ has only two vertices, which belong to every cell, and hence $a \in |\Gamma^k|$ or $\Gamma^k = 0$.

Suppose then that the omission of a principal π of G_1 gives a grating G_0 (itself a refinement of G_σ) for which the theorem is known to be true. Let Q_+, Q_- be the closed half-planes determined by π. If a is a finite point, not in π, let Q be that one of Q_+, Q_- that does not contain a. If $a \in \pi$ or $a = w$, let Q be one of Q_+, Q_-, containing a principal of G_0 parallel to π.

Each k-cell of Γ^k in π belongs to just one $(k+1)$-cell in Q, and therefore if C^{k+1} is the sum of all such $(k+1)$-cells, for all k-cells of Γ^k in π, $\Gamma^k + \dot{C}^{k+1}$ is a k-cycle having no k-cell in π. It follows that *it is the subdivided form on G_1 of a k-cycle Γ_0^k on G_0.* For it could only fail to be so by containing one, but not both, of the halves $A^k Q_+$ and $A^k Q_-$ of a k-cell on G_0. Since these are the only k-cells in G_1 that contain $A^k \pi$ but do not lie in π, it would follow that the $(k-1)$-cell $A^k \pi$ belongs to only one k-cell of $\Gamma^k + \dot{C}^{k+1}$, which is impossible since this chain is a cycle. C^{k+1} *does not contain a.* For if a is finite and not in π, it is not in Q; if $a = w$, all cells of C^{k+1} are finite; if $a \in \pi$ it is not in C^{k+1}, since

$$|C^{k+1}| \pi \subseteq |\Gamma^k|.$$

Thus Γ_0^k is a k-cycle on G_0, not containing a, and so, by the inductive hypothesis, bounds a chain C_0^{k+1} not containing a. Hence, on G_1,

$$(C^{k+1} + C_0^{k+1})^\bullet = \dot{C}^{k+1} + (\dot{C}^{k+1} + \Gamma^k) = \Gamma^k.$$

If $k = p-1$, $C^p + C_0^p$ and its complementary chain are the only chains bounded by Γ^{p-1} (cf. p. 107).

20. Theorem 20·1 (Alexander's Lemma). *Let F_1 and F_2 be closed sets in Z^p, and Γ^k a k-cycle, $(0 \leqslant k \leqslant p-2)$, which, for $i = 1$, 2, bounds a chain C_i^{k+1} in $Z^p - F_i$. Then if $C_1^{k+1} + C_2^{k+1} \sim 0$ in $\mathscr{C}(F_1 F_2)$, $\Gamma^k \sim 0$ in $\mathscr{C}(F_1 \cup F_2)$.*

Let C^{k+2} be a $(k+2)$-chain in $\mathscr{C}(F_1 F_2)$, on a grating **G**, bounded by $C_1^{k+1} + C_2^{k+1}$. The closed sets $|C^{k+2}|F_1$ and $|C^{k+2}|F_2$ do not meet, and therefore a refinement, **G***, of **G** exists, no cell of which meets both these sets. If C_0^{k+2} is the set of $(k+2)$-cells of **G*** that lie in C^{k+2} and meet F_2 it follows, just as in the 2-dimensional case, that the chain

$$\mathring{C}_0^{k+2} + C_1^{k+1},$$

whose boundary is Γ^k, meets neither F_1 nor F_2.

21. A set of points in Z^p which is the homeomorph of a bounded j-cell is called a *j-element*, and denoted by E^j. Since every bounded j-cell can be mapped topologically on to the set

$$\sum_1^j \xi_i^2 \leqslant 1$$

in R^j (Example 4, p. 62), all j-cells are homeomorphic. The homeomorph of the unit j-sphere

$$\sum_1^{j+1} \xi_i^2 = 1 \quad \text{in } R^{j+1} \quad (j \geqslant 0)$$

is called a *topological j-sphere*, and denoted by J^j.

Theorem 21·1. *If E^j is a j-element in Z^p $(0 \leqslant j \leqslant p)$, every k-cycle in $\mathscr{C}E^j$ bounds in $\mathscr{C}E^j$.*

The proof is by induction on j. There is only one non-null p-cycle in Z^p, namely Ω^p, and therefore there is none in $\mathscr{C}E^j$: the highest possible value for k is $p-1$. If $j = 0$, E^0 is a point, and by 19·1 cannot prevent Γ^k from bounding.

Suppose, then, that the theorem is proved for elements of less than j dimensions, for all k less than p, but that Γ^k is a non-bounding cycle in $\mathscr{C}E^j$. Then k cannot be $p-1$, for the continuum E^j cannot meet both the p-chains bounded by Γ^{p-1} without meeting Γ^{p-1}.

Let A^j be the pattern cell of E^j and f a topological mapping of

A^j on to E^j. A^j can be cut into two equal j-cells, A_1^j and A_2^j, with a common $(j-1)$-cell A^{j-1}. The sets

$$E_1^j = f(A_1^j) \quad \text{and} \quad E_2^j = f(A_2^j)$$

are j-elements whose union is E^j and whose common part is the $(j-1)$-element $f(A^{j-1})$. If $\Gamma^k \sim 0$ both in $\mathscr{C}E_1^j$ and in $\mathscr{C}E_2^j$—say bounds C_i^{k+1} in $\mathscr{C}E_i^j$—the cycle $C_1^{k+1} + C_2^{k+1}$ bounds in $Z^p - f(A^{j-1})$, by the inductive hypothesis, and therefore, by Alexander's Lemma, Γ^k bounds in $\mathscr{C}(E_1^j \cup E_2^j)$, i.e. in $\mathscr{C}E^j$, contrary to the assumption. Hence at least one of the halves E_i^j prevents Γ^k from bounding.

The bisection argument used in 10·1 and again in 12·2 can now be applied to shew that there is a point a such that each of a sequence of sets closing down on a prevents Γ^k from bounding. This is impossible, for there is a chain C^{k+1} in $Z^p - (a)$ bounded by Γ^k, and a small enough neighbourhood of a does not meet C^{k+1}. We have thus reached a contradiction.

Theorem 21·2. *If J^j is a topological j-sphere in Z^p, then*

(1) *if $k \neq p - j - 1$ every k-cycle in $\mathscr{C}J^j$ bounds in $\mathscr{C}J^j$;*

(2) *there exists a non-bounding cycle Γ^{p-j-1} in $\mathscr{C}J^j$;*

(3) *if Γ_1^{p-j-1} and Γ_2^{p-j-1} are non-bounding in $\mathscr{C}J^j$ their sum bounds in $\mathscr{C}J^j$.*[19]

The case $j = p - 1$ of this theorem contains the direct generalisation of the classical Jordan's Theorem, namely *a topological $(p-1)$-sphere determines two domains in Z^p*, for (2) asserts the existence of a non-bounding 0-cycle, of which two points must belong to different residual domains, and by (3), if J^{p-1} separates x from y and y from z, it does not separate x from z, i.e. there are at most two domains. An example of the many other results contained in the theorem is: in the residual set of a simple closed curve in Z^3 there is at least one non-bounding 1-cycle.

The proof is by induction on j. If $j = 0$, then the "sphere" J^j is a pair of points, a and b. (1) follows from 19·1 and 20·1, with $F_1 = a$, $F_2 = b$; (2) states that any pair of points prevents some $(p-1)$-cycle from bounding—the boundary of any sufficiently small p-cell with centre one of the points is such a $(p-1)$-cycle; and (3) is the Theorem 6·4, for $(p-1)$-cycles in Z^p, proved exactly as on p. 109. Suppose then that the theorem is completely proved for "spheres" of less than j dimensions.

The unit j-sphere in R^{j+1}, of which J^j is the f-image, is the union of two "hemispheres" lying in $\xi_1 \geqslant 0$ and $\xi_1 \leqslant 0$, whose common part is the $(j-1)$-sphere

$$\sum_2^{j+1} \xi_i^2 = 1, \quad \xi_1 = 0.$$

Thus J^j is the union of two j-elements, E_1^j and E_2^j, whose common part is a topological $(j-1)$-sphere, J^{j-1}.

(1) Let Γ^k be a k-cycle in $\mathscr{C}J^j$, and C_1^{k+1} and C_2^{k+1} the $(k+1)$-chains in $\mathscr{C}E_1^j$ and $\mathscr{C}E_2^j$, bounded by Γ^k, whose existence was proved in 21·1. Since $k+1 \neq p-(j-1)-1$, the cycle $C_1^{k+1}+C_2^{k+1}$ bounds in $\mathscr{C}J^{j-1}$, by the inductive hypothesis. Therefore, by Alexander's Lemma, Γ^k bounds in $\mathscr{C}J^j$.

We next prove (3). Let C_{i1}^{p-j} and C_{i2}^{p-j} be, for $i = 1, 2$, $(p-j)$-chains in $\mathscr{C}E_1^j$ and $\mathscr{C}E_2^j$ respectively, bounded by Γ_i^{p-j-1}. Then $C_{i1}^{p-j}+C_{i2}^{p-j}$ does not bound in $\mathscr{C}J^{j-1}$, for if it did Γ_i^{p-j-1} would bound in $\mathscr{C}J^j$ (by Alexander's Lemma), contrary to hypothesis. Therefore, by the inductive hypothesis, the sum

$$(C_{11}^{p-j}+C_{12}^{p-j})+(C_{21}^{p-j}+C_{22}^{p-j}) \sim 0$$

in $\mathscr{C}J^{j-1}$. But this expression is also the sum of $(C_{11}^{p-j}+C_{21}^{p-j})$ and $(C_{12}^{p-j}+C_{22}^{p-j})$, which are chains, bounded by $\Gamma_1^{p-j-1}+\Gamma_2^{p-j-1}$, in $\mathscr{C}E_1^j$ and $\mathscr{C}E_2^j$ respectively. Hence, by Alexander's Lemma, $\Gamma_1^{p-j-1}+\Gamma_2^{p-j-1} \sim 0$ in $\mathscr{C}J^j$.

(2) By the inductive hypothesis there exists a non-bounding cycle Γ^{p-j} in $\mathscr{C}J^{j-1}$. The two closed sets $E_1^j | \Gamma^{p-j} |$ and $E_2^j | \Gamma^{p-j} |$ do not meet, and there is therefore a refinement, $\mathbf{G}^*$, of the grating of which no cell meets both sets. Let C^{p-j} be the sum of the cells of Γ^{p-j}, considered as a chain on $\mathbf{G}^*$, that meet E_2^j. Then $\dot{C}^{p-j}$ is a non-bounding cycle in $\mathscr{C}J^j$. For if it bounds C_0^{p-j} in $\mathscr{C}J^j$, the cycle $C^{p-j}+C_0^{p-j}$ does not meet E_1^j and therefore bounds in $\mathscr{C}E_1^j$, a fortiori in $\mathscr{C}J^{j-1}$; and $C_0^{p-j}+(C^{p-j}+\Gamma^{p-j})$ does not meet E_2^j and therefore bounds in $\mathscr{C}E_2^j$, a fortiori in $\mathscr{C}J^{j-1}$. Hence the sum of these cycles, namely Γ^{p-j}, bounds in $\mathscr{C}J^{j-1}$, contrary to hypothesis.

This completes the proof of the whole theorem.

One part of the original Jordan's Theorem is not contained in 21·2, namely that J^{p-1} is the frontier of both its residual domains. This may be proved by the methods of part III of Jordan's Theorem in R^2, or as follows. Let a be any point of the unit $(p-1)$-sphere in R^p, and let the points within a small $(p-1)$-sphere with centre a be removed from the unit $(p-1)$-sphere. What remains is a $(p-1)$-element, whose f-image is a $(p-1)$-element, E^{p-1}, contained in J^{p-1} and containing all points of it save those in a small neighbourhood of b, $= f(a)$. Two points, x and y, that are separated by J^{p-1} bound a 1-chain C^1 in $\mathscr{C}E^{p-1}$, but they are not connected in $\mathscr{C}J^{p-1}$ and therefore C^1 meets $J^{p-1} - E^{p-1}$. By a lemma proved in the next chapter,† C^1 contains a simple arc with end-points x and y. The first point of J^{p-1} from x on this arc is a point of closure of the x-domain, and the first from y is a point of closure of the y-domain. Thus an arbitrarily small neighbourhood of b contains points of both domains, and therefore b belongs to the frontiers of both.

From 21·2 there follow, just as in § 3:

Theorem 21·3. (Invariance of dimension number (Brouwer).) R^p cannot be mapped topologically on to R^q unless $p = q$.

Theorem 21·4. (Invariance of open sets (Brouwer).) If an open set in R^p is mapped $(1, 1)$ continuously on to the set E in R^p, E is an open set, and the mapping is a homeomorphism.

The proofs need only trivial modifications.

Theorem 9·2 fails if $p > 2$, but a special case of Alexander's Lemma is still that if F_1 and F_2 are non-intersecting closed sets in Z^p or R^p, and if x and y are connected in $\mathscr{C}F_1$ and $\mathscr{C}F_2$, then they are connected in $\mathscr{C}(F_1 \cup F_2)$. The proofs of the theorems in para. 14 are therefore valid in p dimensions without even verbal alterations.

A "stronger form" of Alexander's Lemma, analogous to 16·1, cannot be proved, or indeed even stated, for general values of k without introducing ideas which are beyond the scope of this book. Only the case $k = p - 2$ goes through just as before, but the corollaries do not follow from this case; and indeed 16·2 and its corollary are not true in R^p if $p \geqslant 3$.

† VI. para. 1.

22. *Jacobian Theorems.* Let $Y_i(\xi_1, \xi_2, ..., \xi_p)$ be, for $i = 1, ..., p$, a continuously differentiable function of the real variables ξ_i, in a domain D of R^p; that is, a function differentiable there, with continuous first-order partial derivatives. We denote by $Y(x)$ the point $(Y_1(x), Y_2(x), ..., Y_p(x))$, by $Y_{ij}(x)$ the derivative $\partial Y_i/\partial \xi_j$, by $j(x)$ the matrix $\{Y_{ij}(x)\}$, and we put

$$J(x) = \det j(x) = \frac{\partial(Y_1, Y_2, ..., Y_p)}{\partial(\xi_1, \xi_2, ..., \xi_p)}.$$

Theorem 22·1.[20] *If $a \in D$ and $J(a) \neq 0$, and if $Y(a) = b$, there exist positive numbers δ, ϵ such that if $y \in U(b, \epsilon)$ the equation $Y(x) = y$ is satisfied by just one point x of $U(a, \delta)$; and x is a continuous function, $X(y)$, of y.*

Let $x, x' \in U(a, \delta) \subseteq D$. Then†

$$(1) \qquad Y_i(x') - Y_i(x) = \sum_{j=1}^{p} (\xi_j' - \xi_j) \, Y_{ij}(u_i),$$

where u_i is on the segment xx', and therefore in $U(a, \delta)$. If δ is small enough, $\det(Y_{ij}(u_i)) \neq 0$, and therefore the equations $Y_i(x') = Y_i(x)$ for all i imply $\xi_j' = \xi_j$ for all j, i.e. $x' = x$. Thus for sufficiently small δ the continuous function $Y(x)$ maps the open set $U(a, \delta)$ (1, 1) on to a set E in R^p containing b. By 21·4, E is open, and so contains a neighbourhood $U(b, \epsilon)$; and $Y(x)$ is a homeomorphism, i.e. its inverse $X(y)$ is continuous.

Theorem 22·1 has the following generalisation. Let z denote the point $(\zeta_1, \zeta_2, ..., \zeta_q)$ of R^q, and (x, z) a point of $R^p \times R^q = R^{p+q}$. Let $Y_i(x, z)$ be, for $i = 1, 2, ..., p$, a real function which is continuous in a domain D of R^{p+q}, and continuously differentiable in the coordinates, ξ_i, of x. The meanings of Y_{ij} and J are as before, but they are functions of the point (x, z) of R^{p+q}.

Theorem 22·2. *If $(a, c) \in D$ and $J(a, c) \neq 0$, and if $Y(a, c) = b$, there exist positive numbers δ and ϵ such that if y in $U(b, \epsilon)$ and z in $U(c, \epsilon)$ are given, the equation $Y(x, z) = y$ is satisfied by just one point x of $U(a, \delta)$; and x is a continuous function, $X(y, z)$, of (y, z).*

As in 22·1, the equation $Y(x', z) = Y(x, z)$ implies, for a given z, that $x' = x$, provided that (x', z) and (x, z) lie in a sufficiently small neighbourhood U_0 of (a, c) in R^{p+q}. Hence the transformation $(x, z) \to (Y(x, z), z)$ maps U_0 continuously and (1, 1) on to a set of points in R^{p+q} containing (b, c), and the result follows as before.

---

† Apply the mean value theorem, for the real variable τ, to
$$Y_i(x(1-\tau) + x'\tau) - Y_i(x).$$

Theorem 22·3. *The coordinates X_i of the functions $X(y)$ (22·1), and $X(y,z)$ (22·2) are continuously differentiable with respect to the coordinates, η_j, of y in a neighbourhood of b, or (b, c).*

Theorem 22·4. *If in 22·2 $Y(x,z)$ is continuously differentiable in ξ_j and ζ_h, the $X_i(y, z)$ are continuously differentiable in η_j and ζ_h, and*

$$\frac{\partial X_j}{\partial \zeta_h} = - \sum_1^p \breve{Y}_{ji} \frac{\partial Y_i}{\partial \zeta_h},$$

where $\{\breve{Y}_{ij}\}$ is the inverse matrix of $\{Y_{ij}\}$.

Theorem 22·5. (Converse Jacobian Theorem.) *If, in some neighbourhood of a, the rank of $j(x)$ has the constant value $r < p$, $p-r$ of the functions Y_i are expressible as continuously differentiable functions of the remaining r in some neighbourhood of a.*

The proofs of these three theorems do not make direct use of topological arguments, and are therefore given in the notes.[20]

Theorem 22·6. *If the functions $Y_i(x)$ are continuously differentiable in D, and $Y(x)$ takes no value twice in D, there is at least one point a in D such that $J(a) \neq 0$.*

If $J(x) = 0$ throughout D let r be its maximum rank in D, attained, say at a. An r-rowed minor of $J(x)$ is non-zero at $x = a$, and therefore throughout some neighbourhood of a contained in D; and by hypothesis $r < p$ and all $(r+1)$-rowed minors vanish in the neighbourhood. Thus the conditions of 22·5 are satisfied: a relation $Y^2(x) = W(Y^1(x))$ holds for all x in some $U(a)$ in D, where for any y of R^p, y^1 and y^2 denote the points $(\eta_1, \eta_2, ..., \eta_r)$ and $(\eta_{r+1}, ..., \eta_p)$ of R^r and R^{p-r}. But by the conditions of the theorem, Y maps $U(a)$ $(1, 1)$ and continuously into R^p, and therefore, by 21·4, topologically on to a neighbourhood $V(b)$, where $b = Y(a)$; and a point y can be found in $V(b)$ not satisfying $y^2 = W(y^1)$. This contradiction shews that the assumption that $J(x) = 0$ throughout D is false.

Chapter VI

SIMPLY- AND MULTIPLY-CONNECTED PLANE DOMAINS

§ 1. SIMPLY-CONNECTED DOMAINS

[*Note.* In §§ 1 and 2 of this chapter the results of Chapter v, §§ 2 to end, are not used except in a few isolated proofs.]

1. The union of a finite set of segments† in R^2 or Z^2 is called a *segmental set of points*. If it is also a simple arc it is a *segmental arc*, if it is a simple closed curve it is a *simple polygon*.

Theorem 1·1. *Any two points a and b of a connected segmental set E in X^2 are connected in E by a segmental arc.*

Let E_0 be a connected segmental set for which the theorem is known to be true, pq any segment meeting E_0 but not contained in it, and suppose $a \in E_0$, $b \in pq - E_0$. There is then a segment bc of pq such that c, but no other point of bc, is in E_0. By hypothesis, a is connected to c by a segmental arc in E_0. Then (arc ac) $\cup bc$ is a segmental arc joining a and b in $E_0 \cup pq$.

From this the theorem follows by induction on the number of segments.

A segmental set is *non-singular* if it is expressible as the union of segments such that (1) the common part of two segments, if not null, is a common end-point of both, (2) no three segments have a common point.

Theorem 1·2. *The components of a non-singular segmental set E are segmental arcs and simple polygons.*

Let ab be one of the segments of E, and E_0 the union of the rest. By an inductive hypothesis, the components of E_0 may be assumed to be segmental arcs and simple polygons. The segment ab meets E_0 either (i) not at all, or (ii) in an end-point, say a, or (iii) in a and b. In case (i) the result is obvious. In case (ii) a is an end-point of an arc-component C of E_0, and $ab \cup C$ is an arc-

---

† A segment in Z^2 can, by definition, contain only finite points.

component of E. In case (iii) a and b may be end-points of an arc-component C of E_0, then $ab \cup C$ is a simple polygon-component; or of different arcs C_1, C_2, then $C_1 \cup ab \cup C_2$ is an arc-component.

Corollary. *A connected non-singular segmental set is a simple polygon or segmental arc.*

2. If a simple polygon (or segmental arc) is the locus of a 1-chain on a grating it is called a *stepped polygon* (or *stepped arc*), and the 1-chain itself is a *polygon-chain* (or *arc-chain*), respectively. The expression "polygon-chain" is often shortened to "polygon", but the distinction between π (polygon-chain) and $|\pi|$ (locus) will be maintained in symbolic relations. This is in line with the conventions of v. para. 3 (and footnote), which remain in force.

A 1-chain is *non-singular* if its locus is non-singular in the sense of para. 1.

Theorem 2·1. *If the 2-chain K is connected and $\overset{*}{K}$ is non-singular, $\mathscr{I} \,|\, K\,|$ is connected.*

If $\mathscr{I} \,|\, K\,|$ is not connected it has two components, D_1 and D_2, whose closures are 2-chains with a common point a but no common edge. This is only possible (as inspection of the finite number of possible cases shews) if there are four 2-cells at a and one is contained in $\overline{D}_1$, the diagonally opposite one in $\overline{D}_2$, and the other two are not in K. But in this case a is a singularity of $\overset{*}{K}$.

Fig. 43

An immediate corollary is

Theorem 2·2. (Jordan's Theorem for a stepped polygon.) *A stepped polygon $|\pi|$ has two complementary domains, of each of which it is the frontier.*

The two 2-chains bounded by the polygon π are connected (v. 3·1), and therefore the interiors are domains with frontier $|\pi|$. Since the union of these two disjoint domains is $Z^2 - |\pi|$ they are its components.

The bounded 2-chain bounded by a polygon-chain π will be called the *inner 2-chain* of π.

3. Let K be a 2-chain on a rectangular grating, G, and let G_1 be a refinement of G such that each edge of G contains at least three edges of G_1. This ensures that cells of G_1 containing different vertices of G do not meet. Let K_1 be the 2-chain formed by adding to K (on G_1) all cells of G_1 that contain singular vertices of $\overset{\bullet}{K}$ and are not already in K. Then K_1 has a non-singular boundary: this may be seen by inspection of the one possible

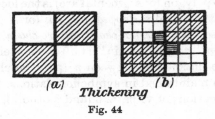

(a) *Thickening* *(b)*

Fig. 44

type of singularity on a rectangular grating (Fig. 44a). K_1 contains K, and each of its components contains one component of K. In particular if K is connected so is K_1.

This process will be called *thickening*.

Theorem 3·1. *If the closed set F in Z^2 is not connected there is a non-bounding stepped simple polygon in $Z^2 - F$.*

Let $H_1 \mid H_2$ be a partition of F, and G a rectangular grating no cell of which meets both H_1 and H_2. Let K, the sum of the 2-cells that meet H_1, be "thickened" into K_1, by means of a refinement G_1 of G of which no cell meets both K and H_2. Then $\overset{\bullet}{K}_1$ is a non-singular 1-cycle which does not bound in $Z^2 - F$; for $\mid K_1 \mid$ contains H_1 and $\mid \Omega^2 + K_1 \mid$ contains H_2. Therefore at least one of the components of $\overset{\bullet}{K}_1$, which are simple polygons, does not bound in $Z^2 - F$ (v. 6·1, Corollary).

Corollary. *Any two disjoint connected closed sets are separated by a stepped simple polygon.*

Theorem 3·2. *If a domain, D, in Z^2 has a connected complement, and if $F \subseteq D$, then F and $\mathscr{C}D$ are separated by a simple polygon.*

Let G be a grating no cell of which meets F and $\mathscr{C}D$, and let K be the sum of the 2-cells meeting F. Form F_0 by joining every pair of components of $\mid K \mid$ by an arc in D, and apply the preceding corollary to F_0 and $\mathscr{C}D$.

Theorem 3·3. *If X and Y are two components of a closed set F in Z^2, there is a simple polygon in $\mathscr{C}F$ separating X and Y.*

Let $H_1 \mid H_2$ be a partition of F such that $X \subseteq H_1$ and $Y \subseteq H_2$ (IV. 5·6). On a grating of which no cell meets both H_1 and H_2 let K be the sum of the 2-cells that meet H_1, and K^* a thickening of K such that $H_1 \subseteq \mathscr{I} \mid K^* \mid$ and $H_2 \subseteq \mathscr{C} \mid K^* \mid$. Then X and Y are separated by $\mid \overset{\ast}{K}{}^* \mid$, and therefore by one of its components.

4. A domain, D, in the closed plane is *simply connected* if every 1-cycle in D bounds in D.†

Example. The following domains in Z^2 are simply connected: the interior of the unit circle; the exterior of the unit circle, i.e. the set

$$[\xi_1^2 + \xi_2^2 > 1] \cup (w);$$

the "punctured plane" $Z^2 - (a)$, where a is any point; the set $\xi_1 > 0$ (without the point at infinity); Z^2 cut along the positive ξ_1-axis, i.e. all points of Z^2 save w and those of

$$[\xi_2 = 0, \ \xi_1 \geqslant 0];$$

the two domains determined by any simple closed curve.

The following domains are not simply connected: $Z^2 - (o \cup w)$; a "punctured circular domain", e.g.

$$[0 < \xi_1^2 + \xi_2^2 < 1];$$

the *finite* points exterior to the unit circle.

A domain in R^2 is said to be simply connected if it is simply connected as a domain in Z^2. Thus $R^2 - (o) = Z^2 - (o \cup w)$ is not simply connected.

The following are necessary and sufficient conditions that a domain D in Z^2 be simply connected:

4·1. *The complementary set $\mathscr{C}D$ is connected.* If $\mathscr{C}D$ is not connected there is a 1-cycle in D which is non-bounding in D (3·1); and if $\mathscr{C}D$ is connected, one of the two 2-chains bounded by any 1-cycle in D contains $\mathscr{C}D$, and the other is in D.

4·2. *Every polygon-chain in D is bounding in D.* The condition is clearly necessary. If D is not simply connected $\mathscr{C}D$ is not connected, and therefore by 3·1 there is a non-bounding polygon-chain in its complement, D.

---

† This definition is suitable only for plane sets. A definition suitable for a large class of spaces is given in Chapter VII.

4·3. *The frontier $\mathscr{F}D$ is connected.* Since every component of $\mathscr{C}D$ contains at least one of $\mathscr{F}D$ (if $\mathscr{F}D \neq 0$)†, the condition is sufficient by 4·1. That it is necessary can be deduced from v. 11·5, Corollary, since $\bar{D}$ and $\mathscr{C}D$ are connected and have the union Z^2, and $\mathscr{F}D$ is their common part. It may also be proved directly as follows. Suppose that, if possible, $\mathscr{C}D$ is connected but $\mathscr{F}D$ is not. Let π be a non-bounding stepped polygon in $Z^2 - \mathscr{F}D$, and let D_0 and D_1 be the residual domains of $|\pi|$. Since the open set D_0 contains a point of $\mathscr{F}D$ it contains points of both D and $\mathscr{C}D$, and similarly these two connected sets both meet D_1. Therefore they both meet $|\pi|$, the common frontier of D_0 and D_1. Thus the continuum $|\pi|$ meets both D and $\mathscr{C}D$, and therefore also $\mathscr{F}D$, contrary to the assumption that it lies in $Z^2 - \mathscr{F}D$.

None of these conditions is sufficient for a domain in R^2 to be simply connected, and only 4·2 is necessary. The complement of a simply connected domain in the open plane may have any number of components (Fig. 45), and on the other hand the set $R^2 - (o)$ is not simply connected. A necessary and sufficient condition that a domain D in R^2 be simply connected is that if γ is any 1-cycle in it, the *finite* 2-chain bounded by γ be contained in D.

Fig. 45

Any of the foregoing properties 4·1, 4·2, 4·3 might equally well have been taken as the definition of a simply connected domain in Z^2, but neither from them nor from the one we have adopted is it clear that simple connection is an intrinsic property of sets, independent of the space in which they are embedded. This independence will appear from another equivalent form of the definition given in Chapter VII.

Theorem 4·4. *Every residual domain of a continuum in Z^2 is simply connected.*

Let D be a residual domain of the continuum F and let π be, if possible, a non-bounding polygon in D. Then both residual domains of $|\pi|$ contain points of $\mathscr{C}D$, and of D, and therefore of $\mathscr{F}D$, i.e. of F. Hence F meets their common frontier $|\pi|$, contrary to the supposition that $|\pi| \subseteq D$.

† If C is a component of $\mathscr{C}D$, by IV. 6·5 $C \cap \mathscr{F}D = \mathscr{F}C \neq 0$, since $C \neq 0, Z^2$.

This theorem, with 4·3, gives an alternative proof of v. 14·4 for the closed plane.

5. Theorem 5·1. *A cross-cut, L, in a simply connected domain D determines two domains in D, both of which are simply connected.*

That $D - L$ has two components (say D_1 and D_2) was proved in v. 11·7. A polygon-chain, π, in D_1 is not prevented from bounding by the continuum $L \cup \mathscr{C}D$ and therefore bounds a 2-chain K in $D - L$. Since K is connected it lies in one component of $D - L$, which must be D_1, the component containing π. Thus $\pi \sim 0$ in D_1.

Theorem 5·2. *If every cross-cut in the domain D destroys its connection D is simply connected.*

Let $H_1 \mid H_2$ be a partition of $\mathscr{C}D$, and δ the distance between H_1 and H_2. If x is a point of D within $\frac{1}{2}\delta$ of a point a of H_1, the nearest point of $\mathscr{C}D$ to x (say u) on the segment ax is within $\frac{1}{2}\delta$ of a and therefore in H_1. Similarly, a point v of H_2 is the end-point of a straight end-cut vy in D. If x and y are joined in D by a segmental arc, λ, and x_1 and y_1 are the nearest points of λ to $\mathscr{C}D$ on ux and vy respectively,

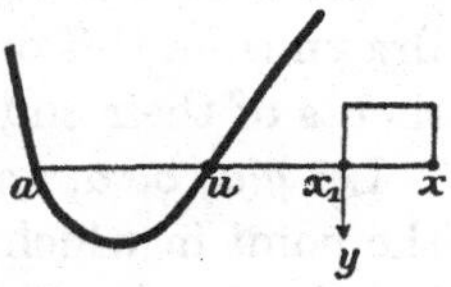

Fig. 46

$$ux_1 \cup \text{arc } x_1 y_1 \cup y_1 v$$

is a cross-cut in D with its end-points on different components of $\mathscr{C}D$. Therefore (v. 16·3) it does not destroy the connection of D.

Theorem 5·3. *If D_1 and D_2 are homeomorphic domains in Z^2, and D_1 is simply connected, D_2 is also.*

Let f be a topological mapping of D_2 on to D_1, and let π be any simple polygon in D_2. Then $f(|\pi|)$ is a simple closed curve in D_1, and one of its complementary domains, say D^*, contains no point of the connected set $\mathscr{C}D_1$. Since f^{-1} maps $\overline{D^*}$ topologically on to a subset of D_1, $f^{-1}(D^*)$ is a domain and $|\pi|$ is its frontier (v. 13·2 and Corollaries 2, 3). Since $\overline{f^{-1}(D^*)}$ is the locus of one of the 2-chains bounded by π, $\pi \sim 0$ in D_2.

(Another proof of 5·3 is contained in Chapter VII.)

§2. MAPPING ON A STANDARD DOMAIN

6. This section is devoted to proving the converse of the theorem at the end of the preceding section, namely that any two simply connected domains in Z^2 (other than the plane Z^2 itself) are homeomorphic. For this purpose it is evidently sufficient to shew that any simply connected domain can be mapped topologically on to some standard domain, e.g. the interior of a circle, or of a square.

Theorem 6·1. *If A is a closed rectangular region, any topological mapping, f, of $\mathscr{F}A$ on to itself can be extended to a topological mapping of A on to itself.* (The meaning is that there is a topological mapping, g, of A on to itself such that if $x \in \mathscr{F}A$, $g(x) = f(x)$.)

Let a be the centre of A. The method is to move the radii drawn from this origin in accordance with the prescribed movements of their end-points on the frontier.

Let $g(a)$ be a, and if x is any other point let y be the point in which the ray ax cuts $\mathscr{F}A$. Then $g(x)$ is defined to be the point dividing the segment $af(y)$ in the ratio ax/ay. The mapping is $(1, 1)$, and if x is on $\mathscr{F}A$, $g(x) = f(x)$; and the continuity of g follows from that of f. Since A is compact it follows that g is a topological mapping.

Fig. 47

Corollary. *If f_1 maps the closed set F topologically on to A, and the subset F_0 on to $\mathscr{F}A$, any other topological mapping, f_2, of F_0 on to $\mathscr{F}A$ can be extended to a topological mapping of F on to A.*

The combined mapping $f_2 f_1^{-1}$ is a topological mapping of $\mathscr{F}A$ on to itself. Let g be a topological mapping of A on to itself coinciding with $f_2 f_1^{-1}$ in $\mathscr{F}A$. Then gf_1 is a topological mapping of F on to A, and if $x \in F_0$

$$gf_1(x) = f_2 f_1^{-1} f_1(x) = f_2(x).$$

Exercise. Extend Theorem 6·1 to any convex closed set in R^p.

Theorem 6·2. *If D is the inner domain of a stepped polygon, $|\pi|$, $\bar{D}$ can be mapped topologically on to a closed rectangular region, A, $|\pi|$ being mapped on to $\mathscr{F}A$.*

(From a corollary to the theorem of the invariance of open sets, v. 13·2, it follows that $|\pi|$ is *necessarily* mapped on to $\mathscr{F}A$; but we do not assume this theorem in proving 6·2.)

The proof is by induction on the number of 2-cells in K, the inner 2-chain of π. That any two rectangular regions are homeomorphic is obvious.

Suppose, then, that K contains more than one 2-cell. If A_0 is a 2-cell of K containing an edge of $\overset{\bullet}{K}$, $|A_0|$ and $|K+A_0|$ have one, two or three edges in common (see Fig. 48); and some subset of these edges is a cross-cut, λ_0, in K, i.e. an arc-chain with its end-points, but no edge, or other vertex, in $\overset{\bullet}{K}$. Let λ_1 and λ_2 be the two arc-chains into which the end-points, x and y, of λ_0 divide π, and let K_1 and K_2 be the inner 2-chains of $\lambda_0+\lambda_1$ and $\lambda_0+\lambda_2$ respectively. *Then* $|K_1|\cup|K_2|=|K|$, *and* $|K_1|\cap|K_2|=\lambda_0$. For since

$$(K_1+K_2)^{\bullet} = \lambda_1+\lambda_2 = \pi,$$

and π bounds only one bounded chain, $K_1+K_2 = K$. If $|K_1|$ contained a point of $\mathscr{C}|K|$ it would contain the whole of this unbounded domain (which contains no point of $\mathscr{F}|K_1|$). Hence $K_1\subseteq K$, and similarly $K_2\subseteq K$. It follows that K_1 and K_2 have no

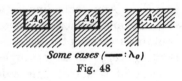

Some cases ($\text{———}:\lambda_0$)

Fig. 48

common 2-cell, for it would be a 2-cell of K not appearing in K_1+K_2. Thus

$$|K_1|\cup|K_2| = |K_1+K_2| = |K|;$$

and $\qquad |K_1|\cap|K_2| = |\overset{\bullet}{K_1}|\cap|\overset{\bullet}{K_2}| = |\lambda_0|.$

Neither K_1 nor K_2 is null, for their boundaries contain λ_0. Therefore each of them contains less 2-cells than K. By an inductive hypothesis their loci can be mapped topologically on to the two halves into which the rectangle A is divided by a segment parallel to one side. Moreover, if $pqrs$ and $pqr's'$ are the two half-rectangles, $|\lambda_0|$, $|\lambda_1|$ and $|\lambda_2|$ can be mapped topologically on to pq, $ps\cup sr\cup rq$ and

Fig. 49

$ps'\cup s'r'\cup r'q$, x falling on p and y on q in each case; and the mappings of $|K_1|$ and $|K_2|$ on the half-rectangles can be made to agree with these mappings of the frontiers. When this is done it is obvious that we have a topological mapping of the whole of $|K|$ on to the whole of A.

Let π_a and π_b be simple polygons, π_a lying in the inner domain of π_b; and let their inner 2-chains be K_a and K_b. If

$$K_0 = K_a + K_b$$

$|K_0|$ is called the *annulus* bounded by the polygons.

Theorem 6·3. *Any annulus can be mapped topologically on to any other.*

Let K_a, K_b and K_0 be as above, and c an interior point of a 2-cell of K_a. Since c is not on the grating a ray from c parallel to the ξ_1-axis does not pass through any vertex. Let x_a be the outer-most intersection of this ray with $|\pi_a|$, and x_b its next intersection with $|\pi_b|$. All points of the segment $x_a x_b$ except x_a and x_b are outside $|\pi_a|$ and inside $|\pi_b|$, and therefore belong to $|K_0|$. Any other line parallel to this ray and sufficiently near to it contains a cross-cut $y_a y_b$ in K_0, such that $x_a x_b y_b y_a$ is a rectangle. The interior of this rectangle evidently contains no point of $|\pi_a|$ or $|\pi_b|$,

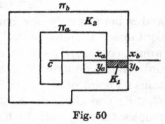

Fig. 50

and, since it contains points outside $|\pi_a|$ and inside $|\pi_b|$, is contained in $|K_0|$. Thus on a suitable refinement $\mathbf{G}_1$ the closed rectangle $x_a x_b y_b y_a$ is a 2-chain, K_1, contained in K_0. If

$$K_2 = K_0 + K_1$$

(i.e. the 2-cells of K_0 not in K_1), K_1 and K_2 have no common 2-cell on $\mathbf{G}_1$, and therefore

$$|K_1| \cap |K_2| = |\mathring{K}_1| \, |\mathring{K}_2| = x_a x_b \cup y_a y_b,$$
$$|K_1| \cup |K_2| = |K_1 + K_2| = |K_0|.$$

Let λ_a and λ_b be the "long" arcs $x_a y_a$ and $x_b y_b$ of π_a and π_b. Then

$$\mathring{K}_2 = \mathring{K}_0 + \mathring{K}_1$$
$$= \lambda_a + \lambda_b + x_a x_b + y_a y_b,$$

a simple polygon.

Let another annulus be exactly similarly treated, the same names being used as before, but with dashes (K'_a, x'_a, etc.). Let the arcs $|\lambda_a|$, $|\lambda_b|$ and the segments $x_a x_b$, $y_a y_b$, $x_a y_a$, $x_b y_b$ be mapped topologically on the corresponding "dashed" arcs, so that correspondingly named end-points are mapped on each other. This gives topological mappings of $|\overset{*}{K}_1|$ and $|\overset{*}{K}_2|$ on $|\overset{*}{K}'_1|$ and $|\overset{*}{K}'_2|$, which agree in the common arcs $x_a x_b$ and $y_a y_b$. Hence if $|K_1|$ is mapped topologically on $|K'_1|$, and $|K_2|$ on $|K'_2|$, by functions agreeing with the given mappings of the boundaries, the whole of $|K_0|$ is thereby mapped topologically on to the whole of $|K'_0|$.

Corollary 1. *The mapping can be chosen so that boundary polygons are mapped on boundary polygons.* This was shewn in the course of the proof. (The boundary polygons *necessarily* correspond, in any mapping, by v. 13·2, but we are not assuming this result.)

Corollary 2. *The mapping can be chosen so as to coincide in $|\pi_a|$ with an assigned topological mapping f of $|\pi_a|$ on to $|\pi'_a|$.* It is evidently sufficient to consider the case where $|K'_0|$ is the annulus bounded by two concentric squares on the grating. The method of the main theorem is followed, but the line $y_a y_b$ is taken so close to $x_a x_b$ that $f(x_a)$ and $f(y_a)$ are on the same edge of the inner square. We may then take x'_a and y'_a to be $f(x_a)$ and $f(y_a)$, and x'_b and y'_b to be the points on π'_b which complete the rectangle. The mapping of $|K_0|$ on to $|K'_0|$ is now completed as before, using f as the mapping of the segment $x_a y_a$ on to the segment $x'_a y'_a$ and of the "long" arc $x_a y_a$ on to the "long" arc $x'_a y'_a$.

Exercise. Shew that the mapping of $|K_0|$ on $|K'_0|$ can be chosen so as to coincide in $|\pi_a|$ with an assigned topological mapping of $|\pi_a|$ on $|\pi'_b|$.

Theorem 6·4. *Every simply connected domain D in Z^2, except Z^2 itself, can be mapped topologically on to the open plane R^2.*

By a preliminary topological mapping of the whole space on to itself it can be arranged that the point at infinity is in $\mathscr{C}D$.

We define inductively a series of rectangular gratings, $\mathbf{G}_n$, and a 2-chain K_n on each $\mathbf{G}_n$. The induction is started by taking

G_0 to be any grating, and K_0 to be the zero-chain. Suppose, then, that for some positive n, G_{n-1} and K_{n-1} have been defined, the latter as a 2-chain on G_{n-1} contained in D. Let G^* be a refinement of G_{n-1} whose finite 2-cells have diameter less than $1/n$ and fill the square

$$E_n: \quad [\,|\xi_1| \leqslant n, |\xi_2| \leqslant n].$$

Let K_n^* be the sum of the 2-cells of G^* that are contained in D. Thus $|K_{n-1}| \subseteq |K_n^*|$. On a suitable refinement, G_n, of G^* there exists a stepped polygon, π_n, such that $|\pi_n|$ separates $\mathscr{C}D$ from $|K_n^*|$ (3·2). (If $K_1^* = 0$ let π_1 be any stepped polygon in D.) We define K_n to be the inner 2-chain of π_n. This completes the inductive definitions.

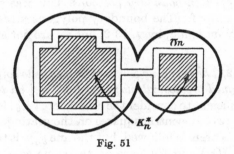

Fig. 51

The chains K_n have evidently the following properties:

(1) $|K_{n-1}| \subseteq |K_n|$,
(2) $\overset{*}{K}_{n-1}$ does not meet $\overset{*}{K}_n$ (for $|K_{n-1}| \subseteq |K_n^*|$),
(3) every point of D belongs to some $|K_n|$.

Hence $|K_n + K_{n+1}|$ is an annulus, Q_n; $Q_n \cap Q_{n+1} = |\pi_n|$, and if $p - q \neq \pm 1$, $Q_p \cap Q_q = 0$.

The method of mapping D topologically on to R^2 is now obvious. We map $|K_1|$, whose frontier is the stepped polygon $|\pi_1|$, on to the square E_1 defined above, $|\pi_1|$ falling on $\mathscr{F}E_1$; Q_1 on to the annulus $\overline{E_2 - E_1}$; and in general Q_n on to $\overline{E_{n+1} - E_n}$. In each case $|\pi_n|$ is mapped on to $\mathscr{F}E_n$, and the mapping of Q_n is to agree on $|\pi_n|$ with the mapping of Q_{n-1}. In this way every point of D receives a unique correlate and every point of R^2 is the image of a point; and the mapping is clearly both ways continuous.

7. It follows from 6·4 that all simply connected domains in Z^2, except the whole closed plane, are homeomorphic. In particular any simply connected domain (with this exception) can be mapped topologically on to the interior of a circle, or on to the open plane R^2. This gives a method of deriving new separation theorems from some of those established in Chapter v. The following theorem is an example.

Theorem 7·1. *If the points x and y of the simply connected domain D of Z^2 are separated in D by F they are separated in D by a component of DF.*

(Two points are *separated in D by E* if they belong to different components of $D - E$.)

Let f map D topologically on to R^2. Since FD is closed in D, $f(FD)$ is a closed set in R^2, separating $f(x)$ and $f(y)$. Hence, by v. 14·3, one of its components, C, separates $f(x)$ and $f(y)$; and $f^{-1}(C)$ is a component of FD separating x and y.

The result of Exercise 2, p. 128, is an immediate corollary of this theorem, and hence (Exercise 3, p. 128) a proof of v. 16·2 is obtained without the "generalised Alexander Lemma".

8. A topological mapping of a simply connected domain D_1 on to another, D_2, cannot in general be extended to a topological mapping of $\overline{D_1}$ on to $\overline{D_2}$, for the frontiers need not be homeomorphic. Even if $\mathscr{F}D_2$ is the homeomorph of $\mathscr{F}D_1$ a given homeomorphism between the domains may not be extensible into a homeomorphism between their closures. Consider, for example, the mapping

$$(r, \theta) \to \left(r, \frac{\theta}{1-r} \right)$$

(r and θ are polar coordinates). This is a topological mapping of the set of points $r < 1$ on to itself, but as the point (r, α) approaches the circumference along the fixed radius $\theta = \alpha$ the image-point describes a curve which approaches the unit circle spirally, and does not converge to any point of it. Clearly no (1, 1)-mapping of the closed circular region on to itself, agreeing with this function in the interior, can be a homeomorphism.

§ 3. CONNECTIVITY OF OPEN SETS

9. Let G be an open set of Z^2, and $k = 0$ or 1. The relation "$\Gamma_1^k \sim \Gamma_2^k$ in G" (read "Γ_1^k is homologous to Γ_2^k in G") was defined in para. 6 of Chapter v as an equivalence relation between k-cycles on the same or different gratings. By it all k-cycles in G, on all possible gratings, are divided into *homology classes*, two cycles belonging to the same class if, and only if, they are homologous in G. The class containing Γ^k will be denoted by $[\Gamma^k]$: thus "$[\Gamma_1^k] = [\Gamma_2^k]$" is another way of writing "$\Gamma_1^k \sim \Gamma_2^k$ in G".

The set of homology classes is made into an abelian group, $H_k(G)$, the kth *homology group*, by defining $[\Gamma_1^k] + [\Gamma_2^k]$ to be $[\Gamma_1^k + \Gamma_2^k]$. The sum is independent of the cycles chosen to represent the classes, for if $\Gamma_1'^k \sim \Gamma_1^k$ in G and $\Gamma_2'^k \sim \Gamma_2^k$ in G, then $\Gamma_1'^k + \Gamma_2'^k \sim \Gamma_1^k + \Gamma_2^k$ in G, i.e. $[\Gamma_1'^k + \Gamma_2'^k] = [\Gamma_1^k + \Gamma_2^k]$. That the addition between elements is associative and commutative follows from the same properties of the addition of cycles; the class $[0]$ containing the zero-chain is a zero of the group, since $[\Gamma^k] + [0] = [\Gamma^k + 0] = [\Gamma^k]$; and every element has an inverse element, namely itself, since $[\Gamma^k] + [\Gamma^k] = [\Gamma^k + \Gamma^k] = [0]$.

Groups in which every element g has period 2 ($g + g = 0$) have particularly simple properties. In each of them there is a *basis*, i.e. a set of generators $\{g_a\}$ which are linearly independent (mod 2). This means that every element of the group is expressible as a sum $g_{a_1} + g_{a_2} + \ldots + g_{a_k}$, but that no such sum is zero if all the g_{a_i} are distinct. In a finite group all bases have the same number, p, of elements, the *rank*, mod 2, of the group. (For an infinite group we shall write $p = \infty$, without distinguishing between different infinite cardinal numbers.)[21]

The rank, $p_k(G)$, of $H_k(G)$, is determined by either of the two properties of being the greatest possible number of k-cycles no linear combination of which bounds in G; or of being the least possible number of k-cycles such that all others are homologous, in G, to a linear combination of them. These properties follow immediately from the corresponding properties of the rank of the group, which is the maximum number of linearly independent elements, and the minimum number of generators. In n-dimensional topology $p_k(G)$ is called the kth Betti number, mod 2; but in the theory of plane sets $p_1(G)$ is usually called simply the *connectivity* of G, and it will soon be seen that no special name is required for $p_0(G)$.

It follows immediately from the remarks just made that a necessary and sufficient condition for a domain D of Z^2 to be simply connected is that $p_1(D) = 0$.

Examples. The domain $[0 < \xi_1^2 + \xi_2^2 < 1]$ has connectivity 1, the domain $Z^2 - (a_1) - (a_2) - \dots - (a_p)$ has connectivity $p - 1$. If F is the union of all circles with centre the origin and radius an integer, closed by the point at infinity, $Z^2 - F$ has infinite connectivity.

10. The notations κ, γ for 1-chains, and K for 2-chains, will now be resumed. Let $c(E)$ be the number of components of any set E.

Theorem 10·1. *If G is a non-null open set in Z^2, $p_0(G) = c(G) - 1$.*

(This includes the assertion that $p_0(G)$ and $c(G)$ are both finite or both infinite.)

First suppose that $c(G)$ is finite. Let $D_1, D_2, \dots, D_c$ be the components of G, and $x_i \in D_i$. If $\Gamma_r^0 = (x_1) + (x_r)$, $r = 2, 3, \dots, c$, the $[\Gamma_r^0]$ *are a basis for* $H_0(G)$. They are linearly independent, for $\Sigma n_r \Gamma_r^0$ ($n_r = 0$ or 1) has a single vertex in any D_r for which $n_r \neq 0$, whereas each component of any 1-chain has an even number of boundary vertices. The $[\Gamma_r^0]$ are generators of $H_0(G)$, for if $x \in D_r$, $(x) \sim (x_r)$ in G, and therefore any 0-cycle Γ^0 is expressible as

$$\Gamma^0 \sim \sum_1^c n_i(x_i) \text{ in } G$$

$$= \sum_2^c n_i \Gamma_r^0 + (x_1) \sum_1^c n_i.$$

Since Γ^0 is a 0-cycle, $\sum_1^c n_i \equiv 0$ mod 2, and therefore

$$\Gamma^0 \sim \sum_2^c n_r \Gamma_r^0 \text{ in } G.$$

If $c(G)$ is infinite the above argument applied to any set of components $D_1, D_2, \dots, D_q$ shews that the corresponding $(x_1) + (x_r)$ are linearly independent, and hence $p_0(G) \geqslant q - 1$ for any q.

6

Theorem 10·2. *The connectivity of an open set G is the sum of the connectivities of its components.*

In each component of G other than the simply connected ones choose a set of 1-cycles that are linearly independent in that component. The whole of these cycles form a linearly independent set in G. For if there were a selection of them bounding the 2-chain K in G, the part of K in any component, D, that it meets would be bounded by a linear combination of the cycles in D, which would therefore be linearly dependent. It follows that if for any component p_1 is infinite, or if there are an infinity of components for which $p_1 > 0$, then (as the theorem requires) $p_1(G)$ is infinite; and that in any case

$$p_1(G) \geqslant \Sigma p_1 \text{ (components)}.$$

If all but a finite number of components are simply connected, and all have finite connectivity, let a maximal linearly independent set of 1-cycles be chosen in each component. *These form a maximal set in G*, for if a further linearly independent cycle γ_0 could be added, the part of γ_0 in at least one component of G would be linearly independent, in that component, of the cycles already chosen there, contrary to the assumption that they are a maximal set.

Thus in all cases $p_1(G)$ is the sum of the connectivities of the components.

Corollary. *If G_1 and G_2 do not meet,*

$$p_1(G_1 \cup G_2) = p_1(G_1) + p_1(G_2).$$

Theorem 10·3. $p_1(G) = c(Z^2 - G) - 1$, *unless $G = Z^2$.*
The case $G = 0$ is trivial, and may be excluded.

Let
$$H_1 \mid H_2 \mid ... \mid H_p$$
be a partition of $\mathscr{C}G$ into p closed, non-intersecting, non-null sets, and on a rectangular grating, $\mathbf{G}$, no cell of which meets two of the sets H_i, let K_i be the set of 2-cells of $\mathbf{G}$ that meet H_i. The linear combination $\gamma = \sum_1^{p-1} n_r \overset{*}{K}_r$ bounds the 2-chains $\Sigma n_r K_r$ and $\Omega^2 + \Sigma n_r K_r$ of which the first meets H_r if $n_r \neq 0$, and the second meets H_p. Thus γ is non-bounding in G unless every $n_r = 0$, i.e. the 1-cycles $\overset{*}{K}_1, \overset{*}{K}_2, ..., \overset{*}{K}_{p-1}$ are linearly independent.

Hence $p_1(G) \geqslant p - 1$. This shews that $p_1(G) \geqslant c(\mathscr{C}G) - 1$ if the latter is finite, and that $p_1(G)$ is infinite if $c(\mathscr{C}G)$ is.

Suppose that $\mathscr{C}G$ has the finite set of components

$$H_1, H_2, \ldots, H_c$$

and let $\mathbf{G}$ and the K_i be as before. Let γ be any 1-cycle in G, and $\mathbf{G}^*$ a refinement of $\mathbf{G}$ such that no cell meets both γ and $\mathscr{C}G$. Let K_i^* be the set of 2-cells of $\mathbf{G}^*$ that meet H_i. The set of points $|K_i^*|$ being the union of all the 2-cells that meet the connected set H_i, is connected, and therefore lies in one of the residual domains of $|\gamma|$. Let K be that 2-chain, bounded by γ, which does not meet K_c^*, and let $n_i = 1$ if $|K_i^*| \subseteq |K|$, and 0 otherwise. Then

$$\gamma + \sum_1^{c-1} n_i \dot{K}_i^* = \left(K + \sum_1^{c-1} n_i K_i^* \right)^{\bullet} \sim 0 \text{ in } G.$$

Now $\dot{K}_i \sim \dot{K}_i^*$ in G, for $K_i + K_i^*$ does not meet H_i, and neither K_i nor K_i^* meets any other set H_j. Hence finally

$$\gamma \sim \sum_1^{c-1} n_i \dot{K}_i \text{ in } G,$$

and therefore $p_1(G) \leqslant c - 1$.

From v. 14·5, it follows that for a domain, D, in Z^2, other than Z^2 itself,

$$p_1(D) = c(\mathscr{F}D) - 1.$$

The connectivity of an open set in the open plane, R^2, is defined to be its connectivity considered as a set in Z^2. Theorem 10·2 remains valid in R^2, but there is no result corresponding to 10·3. (Cf. para. 4.)

Exercise. If $p_1(D)$ is finite there exist $p_1(D)$, but not $p_1(D) + 1$, cross-cuts L_i in D such that $D - \Sigma L_i$ is connected.

11. The theorem that follows is a further example of the use of algebraic methods.

Theorem 11·1. *If D_a and D_b are simply connected domains of Z^2, with complements F_a and F_b, then $c(D_a D_b) = c(F_a F_b)$ if both sides are finite and positive.*

Let $D_1, D_2, \ldots, D_p$ be the components of $D_a D_b$, and let $x_i \in D_i$ for $i = 1, 2, \ldots, p$. Let κ_{ar} and κ_{br} be, for $r = 1, 2, \ldots, p-1$, 1-chains with boundary $(x_r) + (x_p)$, in D_a and D_b respectively.

(1) *The* 1-*cycles* $\gamma_r = \kappa_{ar} + \kappa_{br}$ *are linearly independent in*
$D_a \cup D_b (= Z^2 - F_a F_b)$. If $\sum\limits_{1}^{p-1} n_r \gamma_r \sim 0$ in $D_a \cup D_b$ let Γ^0 be the
common boundary of the 1-chains $\kappa_a = \Sigma n_r \kappa_{ar}$ and $\kappa_b = \Sigma n_r \kappa_{br}$.
Since $\kappa_a + \kappa_b \sim 0$ in $D_a \cup D_b$, $\Gamma^0 \sim 0$ in $D_a D_b$ by Alexander's
Lemma. But $\Gamma^0 = \Sigma n_r ((x_r) + (x_p))$ and has one vertex (only) in
each domain D_r for which $n_r \neq 0$. Thus all the n_r must vanish, and
the γ_r are linearly independent. It follows that

$$c(D_a D_b) - 1 \leqslant p_1(Z^2 - F_a F_b) = c(F_a F_b) - 1.$$

(2) Let H_1, H_2, ..., H_q be the components of $F_a F_b$, and on a
grating **G** of which no cell meets two sets H_i, let K_i be the set of
2-cells meeting H_i. Let **G*** be a refinement of **G** such that no cell
of any subdivided $\overset{*}{K}_i$ meets both F_a and F_b, and let κ_{ci} be the
set of 1-cells of $\overset{*}{K}_i$, on **G***, that meet F_a. *The* 0-*cycles* $\overset{*}{\kappa}_{ci}$, *for*
$i = 1, 2, ..., q-1$, *are linearly independent in* $D_a D_b$. Any linear
combination $\sum\limits_{1}^{q-1} n_i \overset{*}{\kappa}_{ci}$ bounds $\Sigma n_i \kappa_{ci}$ in D_b and $\Sigma n_i (\kappa_{ci} + \overset{*}{K}_i)$ in
D_a. The 1-cycle

$$\Sigma n_i \kappa_{ci} + \Sigma n_i (\kappa_{ci} + \overset{*}{K}_i) = \sum\limits_{1}^{q-1} n_i \overset{*}{K}_i$$

is non-bounding in $D_a \cup D_b$ if some $n_i \neq 0$, and therefore by v. 9·4·1,
$\Sigma n_i \overset{*}{\kappa}_{ci}$ is non-bounding in $D_a D_b$. It follows that

$$c(F_a F_b) - 1 \leqslant p_0(D_a D_b) = c(D_a D_b) - 1.$$

Exercise. Examine the excluded cases, where $c(F_a F_b)$ or $c(D_a D_b)$
is zero, or infinite.

12. Theorem 12·1. $p_1(G)$ *is a opological invariant.*

By 10·2 it is sufficient to prove the theorem for a domain, D.

Let D_1 and D_2 be homeomorphic domains and f a topological
mapping of D_1 on to D_2. Let $x_1, x_2, ...$ be a sequence of points of
D_1 converging to a point of $\mathscr{C} D_1$. The sequence $f(x_1), f(x_2), ...$ in
the compact space Z^2 has a convergent subsequence, say $f(x_{n_r})$,
converging to y. The point y cannot be in D_2, for then (x_{n_r}) would
converge to $f^{-1}(y)$ in D_1.

Let (u_n) be another sequence in D_1 converging to a point of another component of $\mathscr{C}D_1$, and (u_{n_s}) a subsequence such that $f(u_{n_s}) \to v$. *Then y and v are in different components of $\mathscr{C}D_2$.* For let $|\pi|$ be a simple polygon in D_1 separating the components of $\mathscr{C}D_1$ to which (x_n) and (u_n) converge (3·3). For all m and n exceeding a certain n_0, $|\pi|$ separates x_m and u_n.

Now $D_1 - |\pi|$ has two components (v. 11·2) which must each contain one of x_m and u_n, and which are mapped by f on to the two components of $D_2 - f(|\pi|)$. Since y and v are not on $f(|\pi|)$ it follows that they are in different residual domains, and therefore in different components of $\mathscr{C}D_2$.

From this it follows that $p_1(D_1) \leqslant p_1(D_2)$, and similarly $p_1(D_2) \leqslant p_1(D_1)$.

The proof shews (with the help of the "Schröder-Bernstein Theorem", not included in Chapter I) that the cardinal number of components in $\mathscr{C}D_1$ and $\mathscr{C}D_2$, *whether finite or infinite,* is the same —a stronger result than $p_1(D_1) = p_1(D_2)$, with our convention.

Theorem 12·2. *Any two domains in Z^2 with the same finite connectivity are homeomorphic.*

We first shew that if $p_1(D)$ has a finite value, $p-1$, D can be mapped topologically on $Z^2 - (a_1 \cup a_2 \cup \ldots \cup a_p)$, where the a_i's are certain distinct points. It may be assumed that w is in D.

If $H_1, H_2, \ldots, H_p$ are the components of $\mathscr{C}D$, $\mathscr{C}H_1$ is a simply connected domain (IV. 3·3) containing $\overset{p}{\underset{2}{\bigcup}} H_i$, and therefore H_1 can be separated from $\overset{p}{\underset{2}{\bigcup}} H_i$ by a simple polygon, $|\pi_1|$. From the method of constructing this polygon given in the proof of 3·2, it is clear that it can be so chosen that H_1 lies in its inner domain, $\mathscr{I}|K_1|$. The set H_2 can now be similarly separated from

$$|K_1| \cup \overset{p}{\underset{3}{\bigcup}} H_i$$

by a simple polygon $|\pi_2|$. Proceeding in this way we construct p simple polygons, π_i, in D whose inner chains K_i each contain one of the sets H_i, and do not intersect each other.

Let a_i be any point of H_i. The required homeomorphism between D and $Z^2 - \mathsf{U}(a_i)$ is constructed by mapping each point of $|\,\Omega^2 + \Sigma K_i\,|$ on itself, and the set $|\,K_i\,| - H_i$ on $|\,K_i\,| - (a_i)$ by a function leaving the points of $|\,\pi_i\,|$ fixed, constructed in the following way. (Compare throughout the proof of 6·4.)

Since $|\,\pi_i\,|$ is a continuum in the simply connected domain $\mathscr{C}H_i$, there is a polygon $|\,\pi_{i1}\,|$ in $\mathscr{I}\,|\,K_i\,|$ separating $|\,\pi_i\,|$ from H_i, and this polygon may be so chosen that its inner chain K_{i1} is in $U(H_i,\,1)$. We can now construct a simple polygon $|\,\pi_{i2}\,|$ separating $|\,\pi_{i1}\,|$ from H_i and lying in $U(H_i,\,\tfrac{1}{2})$; and so on. Let $\gamma_{i1},\ \gamma_{i2},\ \ldots$ be a sequence of squares, with sides parallel to the coordinate axes, closing down on a_i, such that the largest of

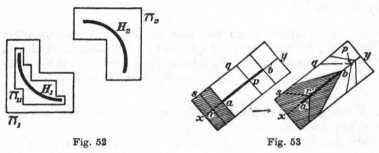

Fig. 52   Fig. 53

them, γ_{i1}, is contained in $\mathscr{I}\,|\,K_i\,|$. Let the annulus bounded by $|\,\pi_i\,|$ and $|\,\pi_{i1}\,|$ be mapped topologically on the annulus bounded by $|\,\pi_i\,|$ and $|\,\gamma_{i1}\,|$ by a function f_1, so that points of $|\,\pi_i\,|$ are mapped on to themselves (6·3); then let the annulus bounded by $|\,\pi_{i1}\,|$ and $|\,\pi_{i2}\,|$ be mapped topologically on that bounded by $|\,\gamma_{i1}\,|$ and $|\,\gamma_{i2}\,|$ by a function agreeing with f_1 on $|\,\pi_{i1}\,|$; and so on. The combination of all these mappings and the identical mapping in $|\,\Omega^2 + \Sigma K_i\,|$ is clearly the required homeomorphism.

It remains to shew that for any two sets of distinct points $a_1, a_2, \ldots, a_p$ and $b_1, b_2, \ldots, b_p$, $Z^2 - \mathsf{U}(a_i)$ and $Z^2 - \mathsf{U}(b_i)$ are homeomorphic, and for this it is sufficient to assume that $a_i = b_i$ except for $i = 1$. Suppose first that the segment $a_1 b_1$ does not contain any other point a_j or b_j. Then the mapping is constructed by leaving fixed all points outside a rectangle with an axis along $a_1 b_1$, small enough to exclude all other points a_j and b_j, and mapping this rectangle on itself in the manner shewn in Fig. 53.

The segment xa_1 is mapped linearly on the segment xb_1, and a_1y linearly on b_1y; and each segment pq running perpendicularly from the line xy to the boundary of the rectangle is mapped linearly on the join of q to the new position, p', of p.

Finally, if a_1b_1 contains a point a_j or b_j we first move a_1 to c, and then from c to b_1, choosing c so that neither a_1c nor cb contains a point a_j or b_j.

This result cannot be extended to general open sets, even if the condition $p_0(G_1) = p_0(G_2)$ is added: see Fig. 54, which shews two sets of three circles in Z^2 whose residual open sets both have connectivity 2 (in accordance with 10·3) but are not homeomorphic. On the other hand, if the components of G_1 and G_2 can be correlated in pairs having the same connectivity, G_1 and G_2 are

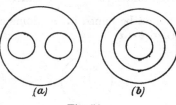

(a)   (b)

Fig. 54

homeomorphic, for the homeomorphisms between the pairs of components together form a homeomorphism between G_1 and G_2.

§4. RELATIONS OF A DOMAIN TO ITS FRONTIER[22]

13. Let E be a subset of a space S, and a any point of S (not necessarily in E). The set E is *locally connected* (lc) at a if, given a positive ϵ, there exists a positive δ such that any two points of $EU(a, \delta)$ are joined by a connected set lying in $EU(a, \epsilon)$. This relation between a and E is independent of the containing space S provided that the metric relations in $a \cup E$ are preserved. This is obvious from the fact that $EU(a, \delta)$ and $EU(a, \epsilon)$ consist of the points of E within δ and ϵ respectively of a.

If $a \in E$ the definition agrees with that already given in Chapter IV. If a is not in $\bar{E}$ then E is certainly lc at a, since for sufficiently small δ there are no points in $EU(a, \delta)$. Thus the interesting new possibilities arise if $a \in \bar{E} - E$, i.e. for an open set G, if $a \in \mathscr{F}G$.

The property is clearly a topological one.

Example. 1. The set of points $[\xi_2 \neq 0]$ in R^2 is not lc at any point of its frontier $[\xi_2 = 0]$; for e.g. the points $(1, \epsilon)$ and $(1, -\epsilon)$, although contained in a 2ϵ-neighbourhood of $(1, 0)$ are not connected in the set. Fig. 55 shews another example: the inner domain is not lc at any point of the "spoke" except its inner extremity.

Let F_1 be the union of the segments joining the origin to the points

$$\left(\tfrac{1}{2}, \tfrac{1}{2}\right), \; \left(\tfrac{1}{2}, \tfrac{1}{3}\right), \; ..., \; \left(\frac{1}{2}, \frac{1}{n}\right), \; ...,$$

and F_2 the boundary of the square $[0 \leqslant \xi_1 \leqslant 1, \; 0 \leqslant \xi_2 \leqslant 1]$. The inner domain of $F_1 \cup F_2$ (Fig. 56) is not lc at $(\alpha, 0)$ if $0 \leqslant \alpha < \tfrac{1}{2}$, for, if $\alpha > 0$, the points $\left(\alpha, \dfrac{4\alpha}{2n+1}\right)$ and $\left(\alpha, \dfrac{4\alpha}{2n-1}\right)$ of the domain (n integral) cannot be joined in the domain by a connected set of diameter less than $\tfrac{1}{2} - \alpha$.

Fig. 55   Fig. 56

A space, or set of points, is *uniformly locally connected (ulc)*, if given a positive ϵ, there exists a positive δ such that all pairs of points, x and y, that satisfy $\rho(x, y) < \delta$ are joined by a connected subset of the space, of diameter less than ϵ.

Example. 2. None of the three sets given in Example 1 is *ulc*: in all three of them there are pairs of arbitrarily close points not joined by a subset of diameter less than $\tfrac{1}{2}$.

All convex domains of R^p are *ulc*, for if $|x - y| < \delta$ the segment xy is a connected subset of the domain, of diameter $< \delta$.

Theorem 13·1. *If E is ulc it is lc at all points of $\bar{E}$; and if E is lc at all points of $\bar{E}$ and $\bar{E}$ is compact, E is ulc.*

(1) Suppose E is *ulc*. For a given ϵ let δ be as in the definition of *ulc*. Then if $a \in \bar{E}$, any two points x, y of $EU(a, \tfrac{1}{2}\delta)$ satisfy $\rho(x, y) < \delta$ and are therefore connected by a subset of E of diameter $< \epsilon$. Since† $\delta < \epsilon$ this subset is contained in $EU(a, 2\epsilon)$.

(2) Suppose E is not *ulc*, and that $\bar{E}$ is compact. Then there is a positive number α, and a sequence of pairs of points x_n, y_n of E such that $\rho(x_n, y_n) \to 0$, but no connected set of diameter less

† Except in the trivial case $E = 0$.

than α joins x_n and y_n in E. The sequence (x_n) in the compact set $\bar{E}$ has a subsequence (x_{n_r}) converging to a point a of $\bar{E}$; and (y_{n_r}) also converges to a. Since any set $EU(a, \delta)$ contains a pair x_{n_r}, y_{n_r}, the set E is not lc at a.

Corollary 1. *In compact spaces, lc implies ulc.*

Corollary 2. *A necessary and sufficient condition for an open set G in a compact lc space to be ulc is that it be lc at all points of $\mathscr{F}G$.* For $\bar{G}$ is compact, and G is lc at its own points.

While it is a topological but relative property of a set to be lc at points of another set, uniform local connection is an intrinsic property but not a topological one. This is shewn by the domains in Figs. 55 and 56 (which are both simply connected, and therefore homeomorphic sets). It is even possible to map the whole of R^2 topologically on itself so that a ulc domain in it is mapped on one that is not ulc (see Fig. 57). But if a compact space is mapped topologically on itself, ulc domains are mapped

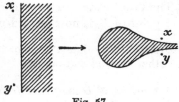

Fig. 57

on ulc domains: this follows from the theorem on uniformity in compact spaces, III. 8·1. The usual process of moving the point at infinity may therefore legitimately be used in dealing with questions of uniform local connection in Z^2. A compact lc space is ulc under all allowed metrics.

14. We now return to the consideration of sets of points in $X^2, = Z^2$ or R^2.

Theorem 14·1. *All Jordan domains are ulc.*

We may assume the Jordan curve, J, to be bounded.

Let a be a point of the curve J, D one of the residual domains, and ϵ a positive number. Let L_1 be an arc of J containing a (not as an end-point), and contained in $U(a, \epsilon)$. If δ is the distance of a from $J - L_1$, any two

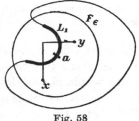

Fig. 58

points, x and y, of D in U (a, δ) are connected in D by a continuum lying in U(a, ε). For if F_ϵ is the frontier of $U(a, \epsilon)$, a path joining x and y in $U(a, \delta)$ meets neither F_ϵ nor $\overline{J - L_1}$, i.e. x and y are not separated by $F_\epsilon \cup \overline{J - L_1}$; and they are not separated by J. Since $J F_\epsilon \subseteq \overline{J - L_1}$,

$$J(F_\epsilon \cup \overline{J - L_1}) = \overline{J - L_1},$$

a continuum, and it follows from v. 9·2 that x and y are not separated by the union, $J \cup F_\epsilon$, of J and $F_\epsilon \cup \overline{J - L_1}$, i.e. x and y bound a 1-chain lying both in D and in $U(a, \epsilon)$.

Theorem 14·2. *The complementary domain of a simple arc, L, in X^2 is lc at the end-points of L.*

The proof exactly reproduces that of 14·1 except that a, the end-point of L, is now an end-point of L_1.

Exercise. Why does the proof break down for a point of a simple arc other than an end-point?

A point of $\mathscr{F}D$ is *accessible from D* if it is an end-point of an end-cut in D.

Example of a domain whose frontier is not everywhere accessible: the domain of Fig. 56. The frontier contains all the points $[\xi_2 = 0, 0 < \xi_1 < \frac{1}{2}]$ and is accessible at none of them. For if the end-cut L has an end-point, a, in this set, there is a subarc ab lying entirely in the domain $|\xi_1| < \frac{1}{2}$. If b is (β_1, β_2), $\beta_1 < \frac{1}{2}$ and $\beta_2 > 0$, and therefore if $1/n < \beta_2$ the segment from o to $(\frac{1}{2}, 1/n)$ cuts the arc ab. $L-(a)$ is therefore not in the given domain.

Exercise. Construct domains whose frontiers have a component (a) containing no accessible point, (b) containing just one accessible point and at least one other point.

Accessibility is clearly a topological property of domain ∪ frontier and a fortiori is invariant under a topological mapping of the whole plane (closed or open) on itself.

The accessible points of the frontier of any domain D are dense in $\mathscr{F}D$. Let a be any point of the frontier and x a point of D in $U(a, \epsilon)$. The nearest point of $\mathscr{F}D$ to x on the segment ax is clearly accessible, and is in $U(a, \epsilon)$.

If a is a finite point, a and x can be joined by an edge or
L-shaped path on a rectangular grating. The nearest point of $\mathscr{F}D$
to x on this path is an end-point of an end-cut in
D which is an edge of a rectangular grating. Such
"directly accessible" points are therefore also
dense in $\mathscr{F}D$.

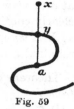

If a is an accessible point of $\mathscr{F}D$ and b a point of
D, there is an end-cut in D joining a to b. For if L_a
is an end-cut with end-points a and x, and L_b a
segmental arc joining x to b in D, and if y is the
first point of L_a, counting from a, that lies in L_b,

Fig. 59

$$(\text{arc } ay \text{ of } L_a) \cup (\text{arc } yb \text{ of } L_b)$$

is the required end-cut. Similarly if a and b are accessible points
of $\mathscr{F}D$ there is a cross-cut ab in D.

Accessibility is a weaker property than relative local connection. The frontier of the bounded domain in Fig. 55 is accessible
at all points, but the residual domain is not lc at points of the
segment, other than the end-point. It will now be shewn that if
D is lc at a of $\mathscr{F}D$, a is accessible from D. The proof depends on:

Theorem 14·3. Let a be a finite point of Z^2 or R^2, and A the union
of a sequence of segments, $x_n y_n$, such that $x_n \to a$, $y_n \to a$. Then if a
is connected in $\bar{A}$ to a point b ($\neq a$), there is a simple arc in $\bar{A}$ with
end-points a and b.

Let C_n be the circle $|x-a| = \dfrac{1}{2n}|b-a|$, for $n = 1, 2, \ldots$.

Since for any n all but a finite number of edges are inside C_n, it
may be arranged (by introducing new end-points) that all intersections of segments with each other and with the circles C_n are
end-points of segments. Thus A_n, the part of A outside or on C_n, is
a segmental set, and therefore, by 1·1, the point b, being connected to a point of C_n by a subset of A_n, is connected to it by
a subarc, J_n, of A_n. Since only a finite number of edges of A lie
outside C_1, an infinity of the arcs J_n have the same initial arc J_1^0
from b to their first point b_1 on C_1. An infinity of these selected
J_n's have (similarly) the same initial arc J_2^0 from b to b_2 on C_2;
and so on. A simple arc J_n^0 is thus defined for every n, joining
b to b_n on C_n, but otherwise lying outside C_n; and $J_n^0 \subset J_{n+1}^0$.

Let ϕ be a topological mapping of the arc $b_n b_{n-1}$ of J_n^0 on to $< 1/2^n,\ 1/2^{n-1} > (b_0 = b)$, and let $\phi(a) = 0$. Then ϕ is a $(1, 1)$ transformation of $X = a \cup \cup J_n^0$ on to $< 0, 1 >$; it is continuous at a in view of the condition $x_n \to a$ for vertices of A; and it is clearly continuous elsewhere. Since the arcs $b_n b_{n-1}$ converge to a, X is compact, and therefore ϕ is topological. Thus X is the required arc.

Theorem 14·4. *If D is lc at a, a point of $\mathscr{F}D$, then a is accessible from D.*

It may be assumed that a is a finite point.

Let δ_n be such that any two points of $DU(a, \delta_n)$ are joined by a stepped arc in $DU(a, 1/n)$. Let x_n be any point of $DU(a, \delta_n)$, and J_n a stepped arc in $DU(a, 1/n)$ joining x_n to x_{n+1}. The set $A = \cup J_n$ has the properties prescribed for A in the preceding Lemma; and if $b \in A$, an arc ab in $\bar{A}$ is an end-cut at a.

The end-cut constructed in this proof has the property that every arc of it not containing a is the locus of a 1-chain on some grating. Such an end-cut is said to be *quasi-linear*.

Corollary. *A simple closed curve in the open or closed plane is accessible at every point from both residual domains.*

This follows at once from 14·1 and 14·4.

Theorem 14·5. *Every simple arc in X^2 is an arc of a simple closed curve in X^2.*

From 14·2 and 14·4 it follows that the end-points of the arc are accessible from the residual domain. Therefore there is a cross-cut joining them, and this with the original arc is a simple closed curve.

Hence *a simple arc is accessible at every point*.

Theorem 14·6. *If L is an arc, with end-points a and b, of the simple closed curve J, and ϵ is any positive number, there is a cross-cut ab, of diameter† less than $\Delta(L) + \epsilon$, in each of the residual domains of J.*

Let ap and bq be end-cuts of diameter less than $\frac{1}{2}\epsilon$ in the residual domain D. Then *p and q are connected in $DU(L, \frac{1}{2}\epsilon)$*. For let L_0 be a closed subarc of $J - L$ such that all points of $J - L_0$ are in

---

† In Z^2 the metric used must be defined throughout a neighbourhood of L.

$U(L, \tfrac{1}{2}\epsilon)$. The points p and q are not separated by $L_0 \cup \mathscr{C}U(L, \tfrac{1}{2}\epsilon)$, for $ap \cup L \cup bq$ is a continuum not meeting this set; and they are also not separated by J. Now

$$J \cap (L_0 \cup \mathscr{C}U(L, \tfrac{1}{2}\epsilon)) = J \cap L_0 = L_0,$$

since $J - L_0$ lies in $U(L, \tfrac{1}{2}\epsilon)$. By v. 9·2, p and q are connected by an arc, L_1, not meeting either set, and therefore lying both in D and in $U(L, \tfrac{1}{2}\epsilon)$.

If x and y are the first points of intersection of ap and bq with L_1 the union of the arcs ax, xy, yb is the required cross-cut.

Since ap and bq may be chosen to be quasi-linear, and L_1 to be segmental, the cross-cut ab may be made quasi-linear.

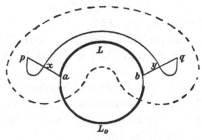

Fig. 60

Exercises. 1. If D is simply connected every cut-point of $\mathscr{C}D$ is accessible from D. [We may suppose $\mathscr{C}D$ is inside E_1, the unit square. For each n form $\mathbf{G}_n$ by dividing E_1 into 4^n equal squares. If $H_1 \mid H_2$ is a partition of $\mathscr{C}D - (a)$ let X be the union of all 2-cells, on all the gratings $\mathbf{G}_n$, that meet H_1 but not H_2. Then $\mathscr{F}X$ is $(a) \cup$ an enumerable set of segments in D. Since $\mathscr{C}D - (a)$ is divided by $\mathscr{F}X$ it is divided by a component of $\mathscr{F}X$, (v. 14·3), and since $\mathscr{C}D$ is connected this can only be the component containing a. Therefore this component $\neq a$, and hence satisfies all the conditions of 14·3.]

2. $\mathscr{F}D$ is said to contain the simple arc L as a *free arc* if no point of L, save possibly the end-points, is a limit-point of $\mathscr{F}D - L$. Shew that L is accessible from D at every point.

15. The following theorem illustrates one of the main causes of the failure of relative local connection of domains.

Theorem 15·1. *A simply connected domain is not lc at cut-points of its frontier.*

Let a be a cut-point and $H_1 \mid H_2$ a partition of $\mathscr{F}D - (a)$. Then $\bar{H}_i = H_i \cup (a)$, (IV. 10·1). If p is any point of H_1 there is an accessible point of $\mathscr{F}D$ nearer to p than any point of $\bar{H}_2$. Thus H_1 contains an accessible point, say b, and similarly H_2 contains an accessible point c.

A cross-cut, L, joining b to c in D determines two domains, D_1 and D_2, in D, and L belongs to both their frontiers (v. 11·7). The point a also belongs to both $\mathscr{F}D_1$ and $\mathscr{F}D_2$. Let $F = \mathscr{F}D_1 \cap \mathscr{F}D$. If a is not in $\mathscr{F}D_1$, $FH_1 \mid FH_2$ is a partition of F and hence the end-points of L are on different components of F. Therefore two points which are not separated by F are not separated by $F \cup L$ (v. 16·3). But

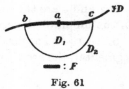

Fig. 61

$F \cup L$ is $\mathscr{F}D_1$, and one of the points may be in D_1 and the other in D_2. Thus the supposition that a is not in $\mathscr{F}D_1$ leads to a contradiction.

Let x and y be points of D_1 and D_2 respectively within a distance $\frac{1}{2}\rho(a, L)$ of a. Then however near to a the points x and y may be an arc joining them in D must cut L and must therefore have diameter greater than $\frac{1}{2}\rho(a, L)$. Therefore D is not lc at a.

Theorem 15·1 remains true without the assumption that D is simply connected.[23]

16. Theorem 16·1.[24] (Converse of Jordan's Theorem, First Form.) *If a closed set has two complementary domains in Z^2, from each of which it is at every point accessible, it is a simple closed curve.*

Let F be the closed set, D_1 and D_2 its residual domains.

F is connected. If it is not, let $\mid \pi \mid$ be a non-bounding simple polygon in $\mathscr{C}F$. If the continuum $\mid \pi \mid$ met both D_1 and D_2 it would meet their common frontier, F, which is not so. Therefore one domain, say D_1, lies in a residual domain of $\mid \pi \mid$. The part of F in the other residual domain of $\mid \pi \mid$ is at a positive distance from D_1, contrary to the accessibility hypothesis.

If a and b are any points of F, $F - (a \cup b)$ is not connected. Let L_1 and L_2 be cross-cuts in D_1 and D_2 connecting a and b (Fig. 62), and let x_1 and x_2 be points of L_1 and L_2 respectively (not a or b).

The simple closed curve $L_1 \cup L_2$ has two residual domains, say D_p and D_q. A cross-cut in D_p from x_1 to x_2 has one end in D_1 and the other in D_2, and therefore meets F. Thus $F - (a \cup b)$ meets D_p, and similarly D_q, but not their common frontier $L_1 \cup L_2$. Therefore it is not connected.

Since F is compact it follows from IV. 12·1 that it is a simple closed curve.

In R^2 the further condition that F be bounded is required.

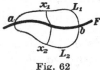

Fig. 62

The following alternative characterisation of a Jordan curve by its relation to the rest of the plane uses the stronger property of uniform local connection but requires it for only one domain.

Theorem 16·2. (Converse of Jordan's Theorem, Second Form.) *If a domain in Z^2 is both simply connected and ulc its frontier is a simple closed curve, or a point, or null.*

Suppose that $\mathscr{F}D$ contains at least two points, a and b. Since D is *ulc* they are accessible and can be joined by a cross-cut, L, which determines two domains, D_1 and D_2, in D. No point of $\mathscr{F}D$ except a and b can belong to the frontier of both these domains, for if c were such a point, any arc in D joining two points x_1 and x_2 of D_1 and D_2 within $\frac{1}{2}\rho(c, L)$ of c would necessarily meet L, and would therefore have diameter at least $\frac{1}{2}\rho(c, L)$. The domain would not be *lc* at c.

Thus the sets

$$E_i = (\mathscr{F}D \cap \bar{D}_i) - (a \cup b) \quad (i = 1, 2)$$

have no common point. Their union is $\mathscr{F}D - (a \cup b)$, and since

$$E_i = (\mathscr{F}D - (a \cup b)) \cap \bar{D}_i$$

they are closed in $\mathscr{F}D - (a \cup b)$. Neither is null, for if

$$\mathscr{F}D \cap \bar{D}_i \subseteq (a) \cup (b)$$

then $\mathscr{F}D_i \subseteq L$, and therefore, by v. 11·7, $\mathscr{F}D_i = L$. This is impossible, since $\mathscr{F}D_i$ separates D_1 and D_2.

Thus $\mathscr{F}D$ is a continuum whose connection is destroyed by the removal of any two points. It is therefore a simple closed curve.

Theorem 16·3. *The components of the frontier of a ulc domain in* Z^2 *are all points or simple closed curves.*

Let D be the domain, F a component of its frontier and F_0 the component of $\mathscr{C}D$ containing F. Then $\mathscr{C}F_0$ is a simply connected domain, D_0 (IV. 3·3), and

$$\mathscr{F}D_0 = \mathscr{F}F_0 = F_0\mathscr{F}D \,(\text{IV. } 6\!\cdot\!4) = F,$$

since F_0 contains only one component of $\mathscr{F}D$ (v. 14·5).

D_0 *is a ulc domain.* Let x be any point of F, ϵ a positive number, and δ such that two points of D within δ of x are joined in D by a path of diameter less than ϵ. Let x_1 and x_2 be points of D_0 within δ of x. If the segment x_1x_2 does not meet F it is a path of diameter less than ϵ in D_0. If it does, and if u_i is the nearest point of intersection to x_i, the segment x_iu_i contains a point v_i

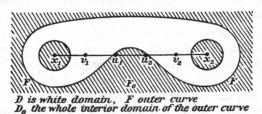

D is white domain, F outer curve
D_0 the whole interior domain of the outer curve

Fig. 63

of D; for otherwise x_iu_i is a continuum joining two components of $\mathscr{C}D$ without meeting D, which is impossible. The points v_1 and v_2 can be joined by a continuum of diameter less than ϵ in D, and the union of this continuum and the segments x_1v_1 and x_2v_2 connects x_1 and x_2 in D_0 and is of diameter less than 3ϵ. Thus D_0 is *lc* at the arbitrary point x of $\mathscr{F}D_0$, and hence is *ulc*. Therefore, by the preceding theorem, its frontier, F, is a point or simple closed curve.

Example. (Kerékjártó's Theorem.)[25] *If two simple closed curves, J_1 and J_2, have more than one common point all the residual domains of $J_1 \cup J_2$ are Jordan domains.*

Clearly none of the residual domains is Z^2 or $Z^2-(a)$, and by v. 14·4 they are all simply connected. It is therefore sufficient to shew that they are *ulc*. We may suppose J_1 and J_2 to be bounded.

Let D be one of the domains, a subset, say, of the residual domains D_1 of J_1 and D_2 of J_2. Let ϵ be a positive number, less than $\Delta(J_1J_2)$, and let δ be such that if x and y are points of D within δ of each other they are connected (for $i = 1, 2$) by a 1-chain κ_i in D_i of diameter less than $\frac{1}{2}\epsilon$. Since $(x) \sim (y)$ in $D_1 D_2$, $\kappa_1 + \kappa_2$ bounds in $\mathscr{C}(J_1 J_2)$, (v. 9·4·1), and therefore, since

$$\Delta(\kappa_1 + \kappa_2) < \Delta(J_1 J_2),$$

$J_1 J_2$ lies in the outer 2-chain of $\kappa_1 + \kappa_2$. Let γ be the boundary of a square on a grating, with centre x and side ϵ. Then κ_1 and κ_2 lie in the inner domain of γ, i.e. γ lies in the outer 2-chain of $\kappa_1 + \kappa_2$. Thus $\kappa_1 + \kappa_2 \sim 0$ in $\mathscr{C}(J_1 J_2 \cup |\gamma|)$. Now κ_i does not meet $J_i \cup |\gamma|$ and the common part of $J_1 \cup |\gamma|$ and $J_2 \cup |\gamma|$ is $J_1 J_2 \cup |\gamma|$. Therefore, by Alexander's Lemma, x is not separated from y by $J_1 \cup J_2 \cup |\gamma|$; i.e. x and y are connected in D and inside $|\gamma|$.

The result is easily extended by induction to any finite number of simple closed curves of which each pair have at least two common points.

§5. TOPOLOGICAL MAPPING OF JORDAN DOMAINS AND THEIR CLOSURES

17. By the following fundamental theorem the topological theory of sets of points in a Jordan domain is reduced to the theory of sets of points in a circular domain.

Theorem 17·1. *If D is a Jordan domain in Z^2, $\bar{D}$ can be mapped topologically on to a closed square region.*

It is sufficient to consider the case where D is bounded, since by a preliminary topological mapping of the whole plane the point at infinity can be placed in the other residual domain.

Let θ_n be the segmental set defined inductively as follows. θ_0 is the frontier of the unit square ($|\xi_1| \leqslant 1, |\xi_2| \leqslant 1$), θ_1 is formed from θ_0 by dividing this square into four equal squares, and θ_n is formed by dividing each of the outermost ring of squares in θ_{n-1} into four equal squares (see Fig. 64).

By using theorems proved in the previous section the process of building up θ_n from θ_{n-1} can be imitated in D. Let f be a topological mapping of θ_0 on to J, the frontier of D. The process of dividing a square into four equal parts can be carried out in the three steps shewn in Fig. 65. This can be reproduced in D as follows. Let a', b', c' and d' be the f-images on J of a, b, c and d on θ_0. Join a' to b' by a quasi-linear cross-cut in D: this determines two domains, D_0 and D_1, in D, each bounded by a simple closed curve (v. 11·8). Since the points c and d belong to the same arc ab of θ_0, c' and d' belong to the same arc $a'b'$ of J—say that belonging to $\mathscr{F}D_0$. Take any point, other than a' or b', of the cross-cut and call it o'. Join o' to c' by a cross-cut in D_0, dividing it into domains D_2 and D_2^*, and let d' be in $\mathscr{F}D_2^*$. This cut can be made segmental except at the end c'. Finally, join o' to d' by a cross-cut in D_2^*, segmental except at the end d', dividing D_2^* into D_3 and D_4.

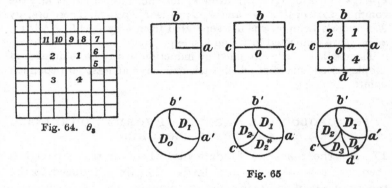

Fig. 64.　θ_2

Fig. 65

The resulting arrangement, ϕ_1, of eight arcs and four inner domains is *isomorphic* with θ_1 and its inner domains in the following sense. The domains, simple arcs, and end-points of simple arcs in the ϕ_1-figure can be put in (1, 1)-correlation with the domains, arcs, and end-points of arcs in the θ_1-figure in such a way that

(1) in both figures two arcs have either a single end-point or nothing in common;

(2) if two ϕ_1-arcs have the point x' in common the correlated θ_1-arcs have the correlated point x in common, and vice versa;

(3) the frontiers of correlated domains are made up of correlated sets of arcs.

N.B. The (1, 1)-correlation is between the domains and arcs *as wholes*: it is not asserted that any correspondence has been set up between their individual points.

Further: (4) correlated arcs of J and θ_0 are also correlates under f;

(5) arcs of ϕ_1 which are end-cuts in D are quasi-linear.

Let us now make the inductive hypothesis that the imitation of the rectilinear figure has been carried as far as θ_{n-1}, i.e. that a system of arcs has been drawn in D such that the resulting figure, ϕ_{n-1}, can be correlated with θ_{n-1} so that (1) to (5) are satisfied, and the following supplement to (5):

(5a) every arc of ϕ_{n-1} which is contained entirely in D is a segmental arc on some grating.

From this assumption it is a simple matter to deduce the existence of ϕ_n having the same relation to θ_n. For θ_n can be derived from θ_{n-1} by quartering the outer squares of θ_{n-1} one by one, in cyclical order; and each square can be quartered by the three steps in Fig. 65. It has already been shewn how this may be imitated in a Jordan domain, and we proceed to carry out this construction in each "cell" of the outer row which, by the inductive hypothesis, exists in ϕ_{n-1}. The choice of the points of division and of the cross-cuts must, however, be further restricted. First suppose that x (see Fig. 66) is not a corner of θ_0. The point p' is to be the f-image of p. If q' and s' are not already fixed from the next cell they are to be chosen so that

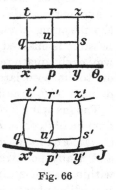

Fig. 66

$$\Delta(\text{arc } x'q') \leqslant \Delta(\text{arc } x'p'),$$
$$\Delta(\text{arc } y's') \leqslant \Delta(\text{arc } y'p').$$

The cross-cut $q'p'$ (the first drawn) is to have diameter less than $3\Delta(\text{arc } x'p')$; u' is to be such that $\Delta(\text{arc } u'p') \leqslant \Delta(\text{arc } y'p')$; and finally the cross-cut $u's'$ is to have diameter less than $4\Delta(\text{arc } y'p')$. Theorem 14·6 ensures that all these conditions can be fulfilled. The point r' and the cross-cut $u'r'$ are not restricted. The cross-cut $q'p'$ is to be a quasi-linear end-cut in D, and $u's'$ and $u'r'$ are to be segmental arcs. A similar procedure is used if x is a corner of θ_0. In this case $q' = f(q)$, and it will easily be verified that the diameters of all arcs in the figure except $z'r'$ and $z's'$ can be made less than the greater of $6\Delta(\text{arc } x'y')$ and $6\Delta(\text{arc } x't')$.

From what has already been said about Fig. 66 it is clear that the resulting figure, ϕ_n, has the required relation to θ_n, and further that the simple closed curves $q'x'p'u'q'$ and $u'p'y's'u'$, and in the case of a corner square, $q'u'r't'q'$ (and therefore also their inner domains) have diameter less than $6\Delta(\text{arc } x'y')$, or when x is a corner of θ_0, less than the greater of $12\Delta(\text{arc } x'y')$ and $12\Delta(\text{arc } x't')$.

Let θ_∞ be the point-set union $\mathsf{U}\,\theta_n$, and let $\phi_\infty = \mathsf{U}\,\phi_n$.[†] All but the outermost ring of squares of θ_n (or ϕ_n) appear undivided and identifiable in θ_∞ (or ϕ_∞). Let the squares of θ_∞ and the corresponding "cells" of ϕ_∞ be numbered, ring by ring outwards from the centre, as indicated in Fig. 64. The cells of the ϕ_∞-figure are bounded by *simple polygons*, each on a certain rectangular grating, and therefore the closure of each cell is the locus of the finite 2-chain bounded by a simple polygon. Map square region 1 topologically on to cell 1 by a function g, which correlates edges and arcs that correspond in the isomorphism between θ_∞ and ϕ_∞. Then map square region 2 on to closed cell 2 by a function which makes correlated edges and arcs correspond, and also agrees with the mapping already set up in the common edge. That this is possible is ensured by 6·2, and 6·1, Cor. By proceeding in this way each square region of the θ_∞-figure is mapped topologically on to its associated ϕ_∞-cell, and every point x of the inner domain of θ_0 is assigned a unique correlate $g(x)$ in D.

The mapping g is evidently a topological mapping of the inner domain of θ_0 on to D. If g is defined to be f in θ_0, g is continuous in *the whole closed square region*. For if p is any boundary point of θ_0 and n any positive integer, the set of all points of the square within a sufficiently small distance of p is contained in at most two adjacent squares of the outer ring in θ_n. Their g-correlates are therefore in two adjacent cells of the outer ring of ϕ_n, i.e. in a set, containing $g(p)$, of diameter not exceeding

$$\eta = 24\Delta(\text{arc } x'y'),$$

where $x'y'$ is the f-image of an edge, xy, of an outermost square of θ_n. The length of xy is $2^{-(n-1)}$, and therefore $\eta \to 0$ as $n \to \infty$. Hence g is continuous throughout the square region, and since it is $(1, 1)$ and the square is compact it follows that g is topological.

It has actually been proved that the mapping of D can be chosen so as to agree with an assigned mapping of J on the

[†] θ_n consists of *edges*, ϕ_n of *arcs*.

boundary of the square. It is also clearly possible to arrange that the point o' which is mapped on to o is an assigned point of D. Hence

Corollary 1. *If J is a simple closed curve in Z^2 it is possible to map Z^2 topologically on to itself so that J is mapped on to the boundary of a 2-cell, A.* It is only necessary to map the closures of the two residual domains of J on to A and $\overline{Z^2 - A}$ by a function which agrees in J with an assigned mapping on A.

Corollary 2. *In Corollary 1, Z^2 may be replaced by R^2.* Map the inner and outer domains of J, considered as a set in Z^2, on to A and $\overline{Z^2 - A}$ by functions which agree along J, and are such that w is mapped on itself.

Corollary 3. *If D_1 and D_2 are Jordan domains in Z^2, a topological mapping f of $\mathscr{F}D_1$ on to $\mathscr{F}D_2$ can be extended to a topological mapping of $\overline{D}_1$ on to $\overline{D}_2$.*

18. Two sets of points, E_1 and E_2, in a space S are said to be *similarly situated* in S if there is a topological mapping of the whole space on to itself which maps E_1 on to E_2. The corollaries 1 and 2 to 17·1 therefore contain the result that *all simple closed curves in the open or closed plane are similarly situated.*

That not all pairs of homeomorphic point-sets in Z^2 are similarly situated is shewn by the sets in Fig. 67. The circles with their projecting rays are evidently homeo-morphic, but it is not possible to map the plane on to itself so that F_1 is mapped on to F_2. For since connection is preserved, each of the residual domains of F_1 would be mapped on to one of the residual domains

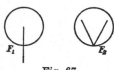

Fig. 67

of F_2, and (since it is a mapping of the whole of Z^2) frontiers on to frontiers. But one of the residual domains of F_2 has a simple closed curve as frontier, whereas the frontiers of the F_1-domains both have cut-points, and are therefore not simple closed curves.

A more interesting example is provided by the simple closed curves of R^3 or Z^3. A simple closed curve in R^3 is *knotted* if it is not similarly situated to a plane circle in the same 3-space, and it is the first step in the theory of knots to shew that, in this sense, knotted curves exist.

Exercises. 1. If L is a simple arc in Z^2 there is a topological mapping of Z^2 on to itself which maps L on a straight segment. [Join the ends of L by a cross-cut in $Z^2 - L$.]

2. If L is a cross-cut in the Jordan domain D, $\bar{D}$ can be mapped topologically on to a circular region so that L is mapped on to a diameter.

3. Prove that if L_1, L_2, ..., L_n are simple arcs in Z^2 which have a common end-point but no other intersection, Z^2 can be mapped topologically on to itself so that the arcs are mapped on to segments radiating from a point. Prove also that in any two such mappings the cyclical order of the image-segments is the same (apart from a possible change of sense).

(This gives a means of defining a cyclical order among the original arcs L_i, and in particular if $n = 4$ of saying which two "separate" the other two.)

Chapter VII

HOMOTOPY PROPERTIES

§ 1. PATHS AND DEFORMATIONS

In the preceding two chapters the properties of plane domains were investigated by the methods of homology, which are based on the addition of k-chains and their boundaries. The methods of the present chapter have a different basis. They are concerned with the extent to which closed paths, or loops, are *homotopic*, that is, deformable into one another within a given set of points. In § 1 the underlying space is for the most part a general metrisable space, § 2 is concerned with the orientation of plane curves.

1. A *path* in a metrisable space S is a mapping $s(\tau)$ of the segment $< 0, 1 >$ into S. The variable τ may conveniently be thought of as time, and it is then in accordance with ordinary usage to regard the path as depending not only on the total set of points passed through, but also on the times at which they are visited. For example, a particle describing the curve in Fig. 68, first in the order *abcdbe*, and then in the order *abdcbe*, is considered to have described two different paths. Two paths s_1 and s_2 are, by definition, different unless $s_1(\tau) = s_2(\tau)$ for every τ in $< 0, 1 >$.

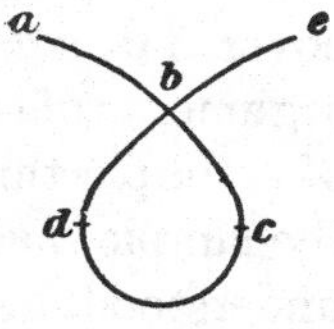

Fig. 68

The set of points in S on to which $< 0, 1 >$ is mapped by s is the *track*, $|s|$. It is compact, and therefore closed in S, connected, and locally connected (IV. 8·2). If $|s|$ is a single point, s is a *point-path*. The points $s(0)$ and $s(1)$ are the *initial* and *final* points of s, and if $0 \leqslant \tau_1 < \tau_2 < \ldots < \tau_k \leqslant 1$, the points

$$s(\tau_1), \ s(\tau_2), \ \ldots, \ s(\tau_k)$$

are by definition in *order* along the path. The path $s(1-\tau)$ is denoted by $-s$; its initial and final points are $s(1)$ and $s(0)$ respectively. If $0 \leqslant \alpha < \beta \leqslant 1$, the *arc* $< \alpha, \beta >$ of s is the path

$s(\alpha + (\beta - \alpha)\tau)$, running from $s(\alpha)$ to $s(\beta)$. If the initial point of a path s_2 is the final point of s_1, $s_1 + s_2$ is the path s_0, where

$$s_0(\tau) = s_1(2\tau) \qquad \text{for } 0 \leqslant \tau \leqslant \tfrac{1}{2}$$
$$\qquad = s_2(2\tau - 1) \quad \text{for } \tfrac{1}{2} \leqslant \tau \leqslant 1.$$

This addition is not associative, since the common end-point of s_1 and s_2 corresponds to the value $\tau = \tfrac{1}{4}$ in $(s_1 + s_2) + s_3$, and to $\tau = \tfrac{1}{2}$ in $s_1 + (s_2 + s_3)$; nor is a path the sum of its arcs $< 0, \alpha >$ and $< \alpha, 1 >$ unless $\alpha = \tfrac{1}{2}$. It will be seen in the next paragraph that these facts cause no inconvenience. We write $s_1 - s_2$ for $s_1 + (-s_2)$ when this path exists.

A *loop* is a path whose end-points coincide: $s(0) = s(1)$. Loops are denoted by the letter l.

In particular examples it is sometimes convenient to allow ranges other than $< 0, 1 >$ for the variable τ. The definitions are easily modified accordingly.

It must be emphasised that paths and loops are not merely sets of points. A concrete illustration, other than the kinematical one given above, is a piece of thread lying in the space, each point of the thread retaining its individuality although several may lie at the same point of the space. The fact that the set of points covered by a path may be very complicated has little effect on the arguments of the present section, just as the complicated shadow of a simple thread lying tangled in a plane has little relation to the simple mechanical properties of the thread. This being well understood, licence will be taken to write "the two paths meet" instead of "the tracks of the two paths meet", etc.

2. *Deformation.* Let $d(\tau, \theta)$ be a mapping of the closed square $[0 \leqslant \tau \leqslant 1, \; 0 \leqslant \theta \leqslant 1]$ of the (τ, θ)-plane (hereafter called "the (τ, θ)-square") into a space S. For each fixed θ, $d(\tau, \theta)$ is a mapping of $< 0, 1 >$ into S, i.e. a path d_θ. Such a series of paths is a *deformation*, and the extreme paths d_0 and d_1 are said to be deformable into one another, in symbols $d_0 \simeq d_1$. The image set of $d(\tau, \theta)$ is the *deformation-set*. It is compact, connected, and locally connected. If it is contained in a subset E of S, then by definition $d_0 \simeq d_1$ *in* E. If d_1 is a point-path, d_0 is *deformable to a point*, and we write $d_0 \simeq 0$.

The process of deformation has a simple interpretation in

terms of the concrete ideas already introduced. If a path or loop is thought of as a piece of thread, and θ is now interpreted as time, the deformation is a continuous movement of the thread as a whole, and $d(\tau, \theta)$, for varying τ and fixed θ, is the position of the thread at time θ. If, on the contrary, τ remains fixed and θ varies, $d(\tau, \theta)$ again determines a path, the *trajectory* of the point (τ) of the piece of thread. An alternative interpretation is to regard the square as a piece of stretchable fabric superposed on S. The θ-paths d_θ are then the segments $\theta = $ const. of the square, in their new positions.

If f is a continuous mapping of the space S_1 into the space S_2, and s is a path in S_1 determined by the function $s(\tau)$, the function $f(s(\tau))$ determines a path in S_2, which is denoted by $f(s)$. If $s_1 \simeq s_2$ in S_1, $f(s_1) \simeq f(s_2)$ in S_2, for if $d(\tau, \theta)$ is the deformation of s_1 into s_2, $fd(\tau, \theta)$ is a deformation of $f(s_1)$ into $f(s_2)$. Similar definitions hold for loops. If $l \simeq 0$ in S_1, $f(l) \simeq 0$ in S_2.

Unless some restriction is imposed on the deformations allowed, the relations $s_1 \simeq s_2$ or $s \simeq 0$ in E are of little interest, since every path s is deformable into a point in its own track $|s|$, by the contraction $s(\tau(1-\theta))$. The two classes of restricted deformation that are of the most interest are as follows.

(1) *Deformation with fixed end-points*: all the paths d_θ have the same initial point a_0 and the same final point a_1, i.e. $d(0,\theta) = a_0$, $d(1, \theta) = a_1$ for all θ. We write $d_0 \cong d_1$.

(2) *Deformation as a loop*: all the paths d_θ are loops, i.e. $d(0, \theta) = d(1, \theta)$ for all θ. This is implied if the letter l is used $(l_0 \simeq l_1$ or $l_0 \cong l_1)$.

If $l \simeq 0$ in E, the loop l is *null-homotopic*.

The relations "$\simeq$ in E" and "$\cong$ in E", as paths or loops, are equivalence relations. For $s \cong s$ by the "identity deformation" $d(\tau, \theta) = s(\tau)$ for all θ; if $s_0 \simeq s_1$ by $d(\tau, \theta)$, then $s_1 \simeq s_0$ by $d(\tau, 1-\theta)$; and if $s_0 \simeq s_1$ by d_1 and $s_1 \simeq s_2$ by d_2, then $s_0 \simeq s_2$ by d_3, where

$$d_3(\tau, \theta) = d_1(\tau, 2\theta) \quad \text{for } 0 \leqslant \theta \leqslant \tfrac{1}{2},$$
$$= d_2(\tau, 2\theta - 1) \quad \text{for } \tfrac{1}{2} \leqslant \theta \leqslant 1.$$

Theorem 2·1. *For any path s, $s - s \cong 0$ as a loop on $|s|$.*

The required deformation has θ-path arc $<0, \theta> -$ arc $<0, \theta>$ of s.

Theorem 2·2. *If $l \simeq 0$ in E then $l \cong 0$ in E.*

Let d be the given loop-deformation, and $\bar{s}$ the trajectory of the end-points in the deformation: $\bar{s}(\tau) = d(0,\tau) = d(1,\tau)$. Let d^* be the deformation with θ-path

$$d^*_\theta = ((\mathrm{arc} < 0,\theta > \text{ of } \bar{s}) + d_\theta) - \mathrm{arc} < 0,\theta > \text{ of } \bar{s}.$$

Thus d^*_θ is d_θ with initial and final "tails" extending along $\bar{s}$ to $l(0)$. Then d^* is a deformation of l into $\bar{s} - \bar{s}$ with fixed end-points. The result now follows from 2·1.

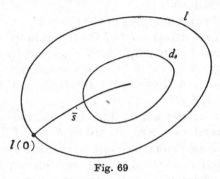

Fig. 69

Theorem 2·3. *If l has its end-points at the final point, a, of s, and $l \simeq 0$ in E, then $s + l \cong s$ in E.*

By 2·2 there is a deformation d^* of l into a with fixed end-points. The required deformation d of $s + l$ into s is given by

$$d(\tau,\theta) = s\left(\frac{2\tau}{1+\theta}\right) \quad \text{for} \quad 0 \leqslant \tau \leqslant \tfrac{1}{2}(1+\theta)$$

$$= d^*\left(\frac{2\tau - (1+\theta)}{1-\theta}, \theta\right) \quad \text{for} \quad \tfrac{1}{2}(1+\theta) \leqslant \tau \leqslant 1.$$

3. It was shewn on p. 62, Example 5, that all topological mappings of $< 0, 1 >$ on to itself are either sense-preserving ("direct"), or sense-reversing ("reversed"), and that the end-points are fixed under direct mappings. Two paths, s_1 and s_2, are *related by change of parameter* if there exists a direct topological mapping, ϕ, of $< 0, 1 >$ on to itself such that $s_2 = s_1\phi$, i.e. $s_2(\tau) = s_1(\phi(\tau))$ for $0 \leqslant \tau \leqslant 1$. Clearly $|s_1| = |s_2|$.

Theorem 3·1. *If s_1 and s_2 are related by a change of parameter,* $s_1 \cong s_2$ *in* $|s_1|$.

Let $s_2 = s_1 \phi$. The function $F(\tau, \theta) = (1 - \theta)\phi(\tau) + \theta\tau$ is a deformation of ϕ into the identity in $<0, 1>$, and for every θ, $F(0, \theta) = 0$, $F(1, \theta) = 1$. Hence $s_1 F$ is the required deformation.

The inconveniences that might be expected to arise from the non-associative addition of paths do not occur in view of this theorem, and the fact that it is *homotopies* (relations $\simeq$, $\cong$) and not identities between paths that are interesting. The sums $s_a = s_1 + (s_2 + s_3)$ and $s_b = (s_1 + s_2) + s_3$ are related by the change of parameter $s_b = s_a \phi$, where

$$\begin{aligned}
\phi(\tau) &= 2\tau && \text{for } 0 \leqslant \tau \leqslant \tfrac{1}{4} \\
&= \tau + \tfrac{1}{4} && \text{for } \tfrac{1}{4} \leqslant \tau \leqslant \tfrac{1}{2} \\
&= \tfrac{1}{2}(\tau + 1) && \text{for } \tfrac{1}{2} \leqslant \tau \leqslant 1
\end{aligned}$$

and therefore $s_1 + (s_2 + s_3) \cong (s_1 + s_2) + s_3$ in its own track. In homotopies we may therefore write $s_1 + s_2 + s_3$ without brackets, and similarly for sums of more than three paths. Again a path s

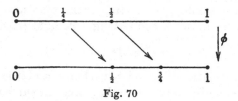

Fig. 70

and the sum, s_σ, of its arcs $<0, \alpha>$ and $<\alpha, 1>$ are not identical (if $\alpha \neq \tfrac{1}{2}$), but they are related by the direct change of parameter $s_\sigma = s\phi$, where

$$\begin{aligned}
\phi(\tau) &= 2\alpha\tau && \text{for } 0 \leqslant \tau \leqslant \tfrac{1}{2} \\
&= \alpha + (1 - \alpha)(2\tau - 1) && \text{for } \tfrac{1}{2} \leqslant \tau \leqslant 1.
\end{aligned}$$

By repeated application of this result it follows that if $0 < \tau_1 < \ldots < \tau_k < 1$, then

$$s \cong \text{arc} <0, \tau_1> + \text{arc} <\tau_1, \tau_2> + \ldots + \text{arc} <\tau_k, 1> \quad \text{of } s,$$

in $|s|$.

s is a *simple path* if $s(\tau) \neq s(\tau')$ whenever $\tau \neq \tau'$; it is a *simple loop* if $s(0) = s(1)$ but $s(\tau) \neq s(\tau')$ whenever $0 < \tau < \tau'$. In order that two simple paths may have the same simple arc as track it is necessary and sufficient that they be related by a change of parameter. If they are directly related they assign the same order to any finite sets of points on the arc. We may therefore speak of a *directed* simple arc, L, and of the order of points along it, when one of the two classes of directly related simple paths on L has been chosen. If l is a loop, and $0 < \alpha < 1$, the function

$$l_\alpha(\tau) = l(\alpha + \tau) \qquad \text{if } \alpha + \tau \leqslant 1$$
$$= l(\alpha + \tau - 1) \quad \text{if } \alpha + \tau > 1$$

is also a loop, with end-points at $l(\alpha)$. It is said to be derived from l by the "rotation α". If l is a simple loop, any other simple loop with the same simple closed curve J as track is derived from l by a rotation followed by a change of parameter. A rotation makes at most a cyclic change in the order along the path of a finite set of points on J.

4. *Paths and loops in R^2 and Z^2.* If s_1 and s_2 are paths in R^2, s_1 may be deformed into s_2 by letting each point $s_1(\tau)$ move with uniform velocity in one second along a straight path to $s_2(\tau)$:

$$s(\tau, \theta) = (1 - \theta) s_1(\tau) + \theta s_2(\tau).$$

s is a loop-deformation if s_1 and s_2 are loops, and it is a deformation with fixed end-points if s_1 and s_2 have the same end-points. This process is called *linear deformation*.

If s_1 and s_2 are paths with the same end-points in the convex set E in R^2, $s_1 \cong s_2$ in E (as loops if they are loops). For the trajectories of the linear deformation are segments joining points of E.

Theorem 4·1. *If s_1 and s_2 are paths with the same end-points in a simply connected domain D in Z^2 or R^2, $s_1 \cong s_2$ in D.*

Let f be a topological mapping of D on to the interior, D_0, of a circle. Then $f(s_1) \cong f(s_2)$ in D_0, and hence $f^{-1}f(s_1) \cong f^{-1}f(s_2)$ in D, i.e. $s_1 \cong s_2$ in D.

Theorem 4·2. *If l is a loop in the simply connected domain D, $l \cong 0$ in D.*

(As 4·1.)

5. If ab is a segment in R^2 or Z^2 we understand by "the path or directed segment ab" the path defined by the mapping

$$a(1-\tau)+b\tau$$

of the segment $<0,1>$. On this understanding "the path $a_1a_2+a_2a_3+\ldots+a_{q-1}a_q$" and "the loop $a_1a_2+a_2a_3+\ldots+a_qa_1$" have well-defined meanings in homotopies. With these definitions, and on the understanding that linear deformations are to be used unless the contrary is stated, geometrical, and even diagrammatic descriptions of paths can be used which are usually shorter and more intelligible than the formula defining the deformation function. For example, the deformation of the two sides $ab+bc$ of a square into the other two sides, $ad+dc$, that is intended in Fig. 71 is clear without any further description.

Fig. 71   Fig. 72

The base of the unit (τ,θ)-square can itself be regarded as a path, from $(0,0)$ to $(0,1)$, and can undergo a deformation d^*

$$d^*(\tau,\theta): \quad (\tau,\theta)\to(\tau',\theta')$$

in the square (Fig. 72). If then $d(\tau,\theta)$ is a deformation in any space S the mapping dd^* of the (τ,θ)-square is also a deformation in S, in which the path $(dd^*)_\theta$ is the image in S of d_θ^* in the square. dd^* is a deformation of s_0 into s_1 which has the same deformation set as d, but may have different intermediate paths. If d is a deformation with fixed end-points, or a loop deformation, so is dd^*. By the use of this fact many properties of deformations in S may be deduced immediately from the corresponding properties in the square.

Theorem 5·1. *Any deformation can be carried out in a finite number of steps, in each of which movement is confined to a small arc of the path.*

The meaning is that in any one step the initial and final paths are of the form

$$u_1 + u_2 + u_3 \quad \text{and} \quad u_1 + u_2' + u_3,$$

where the paths u_1 and u_3 remain fixed and $u_2 \cong u_2'$ in a small neighbourhood. ($u_2 \cong u_2'$ if u_1 or u_3 is absent. The division into arcs u_1, u_2, u_3 is in general different in each step.) We first

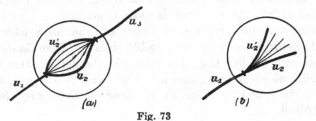

Fig. 73

prove this for the deformation of one side of a square into the opposite side. In Fig. 74 are shewn five stages of a deformation d_r which carries the segment $\theta = (r-1)/m$ of the square into the segment $\theta = r/m$: the sloping portion is supposed to move uniformly from left to right until it finally disappears. By performing in succession the deformations $d_1, d_2, \ldots, d_m$ the lower side

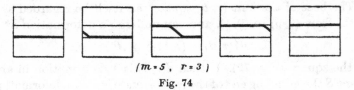

$(m = 5, \quad r = 3)$

Fig. 74

of the square is carried into the upper. This deformation can be split up into partial deformations of the required kind by taking as a "step" the part of d_r in which the sloping segment advances by an amount $1/m$. The whole deformation is thus divided into $m(m+1)$ steps, in each of which movement is restricted to the deformation of two sides, $ab + bc$, of a parallelogram of diameter less than $3/m$ into the other two sides, $ab' + b'c$; and m is at our disposal.

The property now follows immediately for deformations in any space in the manner already explained.

Corollary 1. *If d is a loop-deformation the steps can be made loop-deformations.* For each of the "vertical" sides of the square is mapped on to a point by d.

Corollary 2. *Any deformation in an open set G in R^2 is equivalent to a series of linear deformations in G.*

Exercise. If $s_1 \cong s_2$ in E, $s_1 - s_2 \simeq 0$ in E as a loop; and conversely.

6. Rectifiable paths. Complex integration. Cauchy's Theorem. The arguments that lead from Cauchy's Theorem for a triangle to the "strong" form of the theorem for a general rectifiable simple closed curve are almost entirely topological in character, and form a good example of the application of the methods of topology.

Let z denote either a point of R^2 or its complex coordinate, $\xi_1 + i\xi_2$. We assume as given the definition of integration of continuous functions along straight paths, and

$$\int_{z_1}^{z_2} f(z)\, dz$$

always denotes the integral along the segment $z_1 z_2$. We assume also the usual formal properties of this integral (including the inequality

$$\left| \int_{z_1}^{z_2} f\, dz \right| \leqslant |z_2 - z_1| \cdot \max |f|),$$

and Cauchy's Theorem for triangles, viz. *if the triangular region $z_1 z_2 z_3$ is interior to a domain in which f is regular (holomorphic),*

$$\int_{z_1}^{z_2} + \int_{z_2}^{z_3} + \int_{z_3}^{z_1} f(z)\, dz = 0.$$

Let s be any bounded path lying in a domain, D_f, of regularity of $f(z)$, and let $z_0, z_1, \ldots, z_n$ be points in order along s, z_0 and z_n being the initial and final points of s, and the points being such that each arc $z_r z_{r+1}$ of s has diameter less than the distance of s from $\mathscr{C} D_f$. Then, by definition,

$$\int_s^* f(z)\, dz = \sum_1^n \int_{z_{r-1}}^{z_r} f(z)\, dz.$$

The value of $\int^*$ is independent of the choice of the points z_r, subject

to the condition imposed. For to shew that the values derived from two such sets of points of subdivision are equal it is sufficient to shew that they are equal to the value obtained by combining the two subdivisions, and this will follow by repetition if it is shewn that the introduction of one new point of subdivision, say z_α, between z_r and

z_{r+1} does not alter the value. But the difference between the old value and the new is

$$\int_{z_r}^{z_\alpha} + \int_{z_\alpha}^{z_{r+1}} - \int_{z_r}^{z_{r+1}} f\,dz,$$

i.e. the integral round the triangle $z_r z_\alpha z_{r+1}$, all points of which are within† $\rho(|s|, \mathscr{C}D_f)$ of s, and therefore in D_f. Hence the value is unchanged.

It follows immediately from the definition that

$$\int_{-s}^{*} f\,dz = -\int_{s}^{*} f\,dz, \quad \int_{s_1+s_2}^{*} = \int_{s_1}^{*} + \int_{s_2}^{*} f\,dz,$$

if all the paths are in D and $s_1 + s_2$ exists.

A path s in a metric space is *rectifiable* if the sums $\sum_0^{m-1} \rho(x_i, x_{i+1})$, for all sets $x_0, x_1, ..., x_m$ of points in order along s, have a finite upper bound. The least upper bound is then the *length*, $\lambda(s)$. If s is not rectifiable it is assigned the length "infinity", and "$\lambda(s) < \infty$" may stand for "s is rectifiable". Clearly $\lambda(-s) = \lambda(s)$; if s_0 is an arc of s, $\lambda(s_0) \leqslant \lambda(s)$; and $\lambda(s_1 + s_2) = \lambda(s_1) + \lambda(s_2)$.

If s_1 and s_2 are related by change of parameter, the order of points along them is the same, and $\lambda(s_1) = \lambda(s_2)$. Since all simple paths with the same track L are so related, they all have the same length, which is by definition $\lambda(L)$. On a simple loop a "rotation α" (para. 3) effects at most a cyclic change of order in a set (x_i), provided that they include the old and new end-points, and therefore leads to the same pairs (x_i, x_{i+1}). Thus all simple loops on a simple closed curve J have the same length, which is by definition $\lambda(J)$.

We now return to paths and functions in R^2. The integral

$$\int_s f(z)\,dz$$

is defined, for rectifiable paths and functions continuous on s, to be the limit of the sum

$$\sum_{r=1}^{n} f(z_r)(z_r - z_{r-1})$$

when $\max |z_r - z_{r-1}|$ tends to zero, where $z_0, z_1, ..., z_n$ are points in order along the curve, z_0 and z_n being the end-points. It is shewn in books on Analysis that, in the prescribed conditions, the limit always exists. It is clear from the definition that if l is a rectifiable loop

$$\int_l dz = 0,$$

---

† $\rho(E_1, E_2)$ is the *Euclidean* distance between E_1 and E_2, i.e. the lower bound of $|z_1 - z_2|$ for z_1 in E_1 and z_2 in E_2.

and that
$$\left|\int f\,dz\right| \leqslant \max |f| \cdot \lambda(l).$$

The integral $\int$ is thus defined for a narrower range of paths than $\int^{*}$, but for a wider class of functions, since f need only exist on the path itself, and be continuous there.

Theorem 6·1. *If s is a rectifiable path in the domain, D_f, of regularity of f,*
$$\int_{s}^{*} f(z)\,dz = \int_{s} f(z)\,dz.$$

Let ϵ be any positive number, and $z_0, z_1, \ldots, z_n$ points in order along s (z_0 and z_n being the end-points), such that (1) $\Delta(\operatorname{arc} z_r z_{r+1}) < \rho(s, \mathscr{C}D_f)$, (2) the oscillation of f on the arc $z_r z_{r+1}$ is less than ϵ, and

$$(3) \qquad \left|\int_{s} f(z)\,dz - \sum_{1}^{n}(z_r - z_{r-1})f(z_r)\right| < \epsilon.$$

(That (2) is satisfied if the arcs are sufficiently small follows from III. 8·1, since f is a continuous function of the path variable.)

Then
$$\left|\int_{s}^{*} - \int_{s}\right| \leqslant \left|\sum_{1}^{n}\int_{z_{r-1}}^{z_r} f(z)\,dz - \sum_{1}^{n}f(z_r)\,(z_r - z_{r-1})\right| + \epsilon$$

$$= \left|\sum_{1}^{n}\int_{z_{r-1}}^{z_r}\{f(z) - f(z_r)\}\,dz\right| + \epsilon$$

$$\leqslant \epsilon\sum_{1}^{n}|z_r - z_{r-1}| + \epsilon \leqslant \epsilon(\lambda(s) + 1).$$

Since ϵ is arbitrary the integrals are equal.

This result shews that $\int$ and $\int^{*}$ are equal whenever both are defined, and the symbol $\int^{*}$ may therefore be dropped.

Theorem 6·2. *If $s_1 \cong s_2$ in D_f,*
$$\int_{s_1} f\,dz = \int_{s_2} f\,dz.$$

The deformation can be carried out in a succession of steps in each of which movement is confined to an arc contained in a neighbourhood, in D_f, of diameter $\rho(|s|, \mathscr{C}D_f)$. On the initial and final paths of such a

7

"step" the points z_r used in defining $\int^*$ may evidently be so chosen that they do not move in the deformation, and the integrals $\int^*$ are then identical for the two paths. Thus the integral $\int^*$ is unaffected by a single step, and therefore by the complete deformation.

Theorem 6·3. *If* $l \simeq 0$ *in* D_f,

$$\int_l f \, dz = 0.$$

This follows immediately from 6·2 and 2·2.

Theorem 6·4. *If* f *is regular in the simply connected domain* D_f, *and* l *is a loop in* D_f, $\int_l f \, dz = 0$. (Follows immediately from 6·3 and 4·2.)

This theorem is to be distinguished from Cauchy's Theorem, which (in its weaker form), however, is an easy deduction from it.

Theorem 6·5. (Cauchy's Theorem, weaker form.) *If* $|l|$ *is a simple closed curve, and if* $|l|$ *and its inner domain are contained in* D_f, *a domain of regularity of* f, *then*

$$\int_l f \, dz = 0.$$

Let D_1 be the inner domain of $|l|$. Then $\mathscr{C}\overline{D_1}$ is a simply connected domain, and $\overline{D_1}$ can be separated from $\mathscr{C}D_f$ by a simple polygon (VI. 3·2), whose inner domain is a simply connected domain contained in D_f and containing l. The result therefore follows from 6·4.

7. The form of Cauchy's Theorem proved above is sufficient for the majority of applications in ordinary analysis, but it is a "weak" form in the sense that it assumes a larger domain of existence and regularity for f than is necessary. The proof of the "stronger" form of the theorem depends on the following lemma on rectifiable paths. (It can be shewn, by more elaborate methods, that the value $5\lambda_0$ can be improved to the "best possible" value $\lambda_0 + \epsilon$ for arbitrary positive ϵ.[26])

Theorem 7·1. *Let* J *be a simple closed curve, and* L_1 *a rectifiable arc of* J, *of length* λ_0, *whose end-points,* a *and* b, *are directly accessible† from the inner domain,* D, *of* J. *Then there is a segmental cross-cut from* a *to* b *in* D *of length* $< 5\lambda_0$.

† a is *directly accessible* from D if there is a straight end-cut ap in D parallel to a coordinate axis.

Let ap, bq be linear end-cuts, each parallel to a coordinate axis. Join p and q by a stepped path κ in D. Divide L_1 into n arcs of equal length $\epsilon < \frac{1}{2}\rho(L_1, |\kappa|)$. Let Q_i be a square region with centre at the mid-point† of arc $x_i x_{i+1}$, and sides $\epsilon + \eta$ (where $\epsilon > \eta > 0$) parallel to the axes. Then $X = \cup Q_i$ does not meet κ, and $L_1 \subseteq \mathscr{I}X$. The boundary of X can be made non-singular by "thickening" (VI.3) without spoiling these properties, and without changing its length. Thus L_1 and κ are separated by a component of the new boundary, i.e. by a stepped polygon, π, and if η is suitably chosen

$$\lambda(\pi) \leqslant 4n(\epsilon + \eta) = 4\lambda_0 + 4n\eta < 5\lambda_0.$$

The Q_i may be chosen so that π has general intersection with ap and bq.

There is an arc of π in $\bar{D}$ meeting both ap and bq. Since a is inside and p outside π, ap has an odd number of intersections with π (v. 8·1) and hence with at least one component, Y, of $|\pi|\bar{D}$. Let L_2 be an arc of $J - L_1$ such that $J - L_2$ is in the inner domain of π (Fig. 75). $Y \cup L_2$ does not separate a and b, since they are joined by L_1, not meeting Y or L_2. Therefore (v. 8·2) the 1-chain $ap + \kappa + qb$ meets Y an even number of times. Since Y does not meet κ, and meets ap an odd number of times, it must also meet bq. If u and v are the nearest points of Y to a and b respectively, $au \cup (\text{arc } uv \text{ of } Y) \cup vb$ is the required cross-cut.

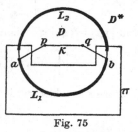

Fig. 75

Theorem 7·2. (Cauchy's Theorem. Stronger form.)[27] *If $f(z)$ is regular in the inner domain, D, of a rectifiable simple closed curve J, and continuous in $\bar{D}$, then*

$$\int_J f(z)\, dz = 0.$$

Let ϵ be any positive number, and δ such that if z_1 and z_2 are points of $\bar{D}$ for which $|z_1 - z_2| < \delta$, then $|f(z_1) - f(z_2)| < \epsilon$. Let $z_1, z_2, \ldots, z_n$ be directly accessible points in order round J such that

$$\lambda(\text{arc } z_r z_{r+1}) < \frac{\delta}{5} \quad (r = 1, \ldots, n;\ z_{n+1} = z_1),$$

and let s_r be the cross-cut in D, with end-points z_r and z_{r+1} of length not exceeding δ, whose existence has just been proved. (s_r also denotes the *path* from z_r to z_{r+1} along this arc.) From the method of proving

---

† I.e. the point bisecting the length of the arc.

7·1 it is clear that for each r the paths can be chosen so that s_{r-1} and s_r have a segment issuing from z_r in common. Therefore

$$\sum_1^n \int_{s_r} f(z)\,dz = \int_l f(z)\,dz,$$

where l is a loop lying entirely in D. Hence this integral is zero.

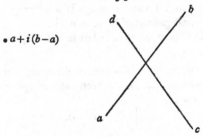

Fig. 76

But if t_r is the arc $z_r z_{r+1}$ of J (considered as a path)

$$\left| \int_J f\,dz - \sum_1^n \int_{s_r} f\,dz \right| \leqslant \sum_1^n \left| \int_{t_r} f\,dz - \int_{s_r} f\,dz \right|$$

$$\leqslant \sum_1^n \left| \int_{t_r - s_r} f(z_r)\,dz \right| + \sum_1^n \left| \int_{t_r - s_r} \{f(z) - f(z_r)\}\,dz \right|.$$

Now

$$\int_{t_r - s_r} f(z_r)\,dz = f(z_r) \int_{t_r - s_r} dz = 0,$$

and $|f(z) - f(z_r)| < \epsilon$ when z is on t_r or s_r. Therefore

$$\left| \int_J f\,dz \right| \leqslant \sum_1^n \epsilon\,\lambda(t_r - s_r) \leqslant \epsilon \cdot 6\Sigma\lambda(t_r) = 6\epsilon\lambda(J).$$

Since ϵ is arbitrary, it follows that $\displaystyle\int_J f(z)\,dz = 0$.

Fig. 77 $\nu(ab,\ cd) = 1$

§ 2. INTERSECTION AND ORIENTATION OF PATHS IN R^2

8. In this section the point (ξ_1, ξ_2) of R^2 is regarded as the complex number $\xi_1 + i\xi_2$. The *left* and *right sides* of the directed segment ab (the path $s(\tau) = a + \tau(b-a)$, $0 \leqslant \tau \leqslant 1$) are defined to be the residual domains of the line ab containing the points $a + i(b-a)$ and $a - i(b-a)$ respectively. A directed segment cd *crosses* a directed segment ab *positively* if it meets it in a point

other than a or b, and if c is on the right side of ab and d on the left; and dc then crosses ab negatively. In these circumstances the *intersection-number*, or Kronecker index, $\nu(ab, cd)$ is defined to be 1, and $\nu(ab, dc) = -1$. If ab and cd do not meet, $\nu(ab, cd) = 0$.

If the segmental paths σ_1 and σ_2 have general intersection, i.e. if no vertex of one lies on the other, $\nu(\sigma_1, \sigma_2)$ is defined to be the excess of positive over negative crossings of σ_1 by σ_2:

$$\nu(\sigma_1, \sigma_2) = \Sigma\nu(a_r a_{r+1}, b_s b_{s+1}) \quad \text{if} \quad \sigma_1 = \Sigma a_r a_{r+1}, \sigma_2 = \Sigma b_s b_{s+1}.$$

It follows immediately from these definitions that

(A) $\nu(-\sigma_1, \sigma_2) = \nu(\sigma_1, -\sigma_2) = -\nu(\sigma_1, \sigma_2),$

(B) $\nu(\sigma_1, \sigma_2 + \sigma_3) = \nu(\sigma_1, \sigma_2) + \nu(\sigma_1, \sigma_3),$
    $\nu(\sigma_1 + \sigma_2, \sigma_3) = \nu(\sigma_1, \sigma_3) + \nu(\sigma_2, \sigma_3).$

An elementary calculation shews† that $\nu(ab, cd) = -\nu(cd, ab)$ and hence

(C) $\nu(\sigma_1, \sigma_2) = -\nu(\sigma_2, \sigma_1).$

Theorem 8·1. *If a segment ab and a segmental loop σ have general intersection, and all intersections of σ with the line ab are in the segment ab, then $\nu(ab, \sigma) = 0$.* For crossings from left to right and from right to left must alternate along σ, and be equal in number.

A segmental path $\sigma = x_0 x_1 + \ldots + x_{k-1} x_k$ is an ϵ-*approximation* to a path s if there exist points y_r in order along s such that (1) (arc $y_r y_{r+1}$ of s) $\subseteq U(x_r, \epsilon)$ for each r, and (2) $x_0 = y_0 = s(0)$, $x_k = y_k = s(1)$. For any two paths s_1, s_2 let $\eta(s_1, s_2)$ be the least distance of the end-points of one path from the track of the other. Thus $\eta(s_1, s_2) > 0$ is equivalent to (and a convenient abbreviation for) the statement that neither path contains an end-point of the other.

Theorem 8·2. *Given a path s and a directed segment ab, $\nu(ab, \sigma)$ has the same value for all $\frac{1}{3}\eta$-approximations, σ, to s that have general intersection with ab, where $\eta = \eta(ab, s)$.*

The theorem is vacuous unless $\eta > 0$. From any two $\frac{1}{3}\eta$-approximations, σ and σ', to s a combined approximation σ'' is

---

† If $\nu(ab, cd) = \pm 1$ it has the same sign as the imaginary part of $(d-c)/(b-a)$.

obtained by joining successive vertices (x_r) and (x'_s) by segments, in the order of the combined sets (y_r) and (y'_s) along s. We can pass from σ to σ' by inserting the new points one at a time, and then removing the old ones, and all the intermediate paths are $\frac{1}{3}\eta$-approximations. It is therefore sufficient to consider the effect of inserting one new vertex x' on σ, the corresponding y', lying (say) on the arc $y_r y_{r+1}$ of s. The segment $x_r x_{r+1}$ of σ is replaced by $x_r x' + x' x_{r+1}$. The sides of the triangle $x_r x_{r+1} x'$ that contain x_r are of length $< \frac{2}{3}\eta$ (see Fig. 78). Therefore the triangle lies in $U(x_r, \frac{2}{3}\eta) \subseteq U(y_r, \eta)$, and so cannot contain a or b. Hence, by 8·1, $\nu(ab, x_r x_{r+1} x') = 0$. The contributions of $x_r x_{r+1}$ and of

$$x_r x' + x' x_{r+1}$$

to ν are therefore equal, i.e. ν is unaltered by the insertion of x'.

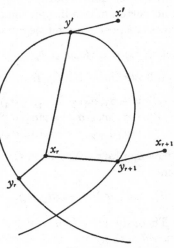

Fig. 78

If $\eta(s, ab) > 0$ a $\frac{1}{3}\eta$-approximation, σ, can always be found having general intersection with ab. The common value of $\nu(ab, \sigma)$ for all such σ is called the *intersection number*, or Kronecker index, $\nu(ab, s)$; and if σ is $a_0 a_1 + \ldots + a_{k-1} a_k$, $\nu(\sigma, s) = \Sigma(a_r a_{r+1}, s)$, provided that $\eta(a_r a_{r+1}, s) > 0$ for each r. The intersection number gives precise expression to the intuitive notion of "the number of times s crosses ab from right to left". The properties (A) and (B) follow immediately for the more general intersections.

Note. If the segmental paths σ_1 and σ_2 have general intersection the number $\nu(\sigma_1, \sigma_2)$ has now been given two definitions, but both give the same value, since the introduction of new vertices y_r along σ_i gives approximations of arbitrary fineness, on taking x_r to be y_r.

Theorem 8·3. *If $\eta(\sigma_0, s) > 0$, and if $\sigma_0 \cong \sigma_1$ without touching $s(0)$ or $s(1)$, then $\nu(\sigma_0, s) = \nu(\sigma_1, s)$.*

We may suppose the deformation to be a combination of

linear ones, and hence that each intermediate path σ_θ is segmental. Let η_0 be the (positive) *g.l.b.* of $\eta(\sigma_\theta, s)$. For some θ_0
in $<0,1>$ let σ be a $\frac{1}{3}\eta_0$-approximation to s, having general
intersection with σ_{θ_0}. If $|\theta - \theta_0|$ is small enough, σ_θ crosses the
same segments of σ as σ_{θ_0}, each in the same direction, and hence
$\nu(\sigma_\theta, s) = \nu(\sigma_{\theta_0}, s)$. Thus $\nu(\sigma_\theta, s)$ is a continuous function of θ, and
since it takes only integral values it is a constant.

Theorem 8·4. *Let l be a loop, and a and b points not on l. Then
$\nu(\sigma, l)$ has the same value for all segmental paths σ from a to b.*

Let σ_0 and σ_1 be two such paths, and d any deformation of σ_0
into σ_1 with fixed end-points. Let s be any path, not passing
through a or b, running from $l(0)$ to a point not in the deformation
set $|d|$. Then $\nu(\sigma_i, -s+l+s) = \nu(\sigma_i, l)$ for $i = 0, 1$ and by 8·3
$\nu(\sigma_0, -s+l+s) = \nu(\sigma_1, -s+l+s)$.

Let the number thus shewn to depend only on l and the points
a and b be denoted temporarily by $\phi(a, b, l)$.

Corollary to 8·4. $\phi(a, b, l)$ *remains constant if a or b moves in
a residual domain of l.* For if (e.g.) σ' joins the point a' to a in
$\mathscr{C}|l|$, $\qquad \phi(a', b, l) = \nu(\sigma' + \sigma, l) = \nu(\sigma, l) = \phi(a, b, l).$

The *order*, $\omega(a, l)$, of a for l is defined, for any loop l and point
a not on l, to be $\phi(a, z, l)$, where z is any point of D_i^∞, the unbounded
residual domain of l. Thus $\omega(a, l)$ is the algebraic number of times
that l crosses any ray (half-line) issuing from a, and hence gives
exact expression to the intuitive notion of "the number of times
that l winds counter-clockwise round a". Clearly

$$\omega(a, -l) = -\omega(a, l), \quad \omega(a, l_1+l_2) = \omega(a, l_1) + \omega(a, l_2),$$

and by 8·4, Corollary, $\omega(a, l)$ remains constant when a moves in
a residual domain of l. In D_i^∞, $\omega(a, l) = 0$.

Theorem 8·5. *If $a, b \in \mathscr{C}|l|$, and σ joins a to b,*

$$\phi(a, b, l) = \nu(\sigma, l) = \omega(a, l) - \omega(b, l).$$

For if σ_a and σ_b join a and b respectively to z of D_i^∞,

$$\nu(\sigma, l) = \nu(\sigma_a - \sigma_b, l) = \nu(\sigma_a, l) - \nu(\sigma_b, l) = \omega(a, l) - \omega(b, l).$$

The notation $\phi(a, b, l)$ is thus no longer needed.

Corollary. *If the segmental path σ_1 has general intersection with the segmental loop σ_2, and crosses it an odd number of times (without regard to sign), its end-points are separated by σ_2.*

Compare v. 8·1. A theorem analogous to v. 8·2 may be deduced as in v. 8.

If m is an integer, the loop $s(\tau) = e^{2m\pi i\tau}$ (the unit circle described $|m|$ times in one or other direction) is denoted by $\mathbf{c}_m$. If $m = 0$, $\mathbf{c}_m$ is the point-path $s(\tau) = 1$. *The order $\omega(o, \mathbf{c}_m)$ is m.* This is obvious if $m = 0$. If not let b be the point (-2) of the negative real axis. Then $\eta = \eta(ob, \mathbf{c}_m) = 1$, and a $\frac{1}{3}\eta$-approximation to $\mathbf{c}_m$ is given by $x_r = y_r = s(\tau_r), \tau_r = r/19\,|m|$ $(0 \leqslant r \leqslant 19\,|m|)$. The segment ob is crossed by those $|m|$ segments $x_r x_{r+1}$ for which $r \equiv 9 \pmod{19}$, positively or negatively as $m \gtrless 0$.

Theorem 8·6. *If $l_1 \simeq l_2$ in $R^2 - (a)$ then $\omega(a, l_1) = \omega(a, l_2)$.*

The deformation can be carried out in a series of steps, d_r, in each of which movement is confined to a domain D_r, of diameter less than the distance of a from the deformation set. If l'_r, l'_{r+1} are the initial and final loops of step d_r, and if σ_r joins a to a point z of the outer domain without meeting D_r, then

$$\nu(\sigma_r, l'_r) = \nu(\sigma_r, l'_{r+1}),$$

i.e. $\omega(a, l'_r) = \omega(a, l'_{r+1})$. From this the result follows.

Corollary 1. *If $l \simeq 0$ in $R^2 - (a)$, $\omega(a, l) = 0$.*

Corollary 2. *If $l_1 \simeq l_2$ in $R^2 - (\sigma(0) \cup \sigma(1))$ then $\nu(\sigma, l_1) = \nu(\sigma, l_2)$.* (By 8·5.)

This result can be used to extend the definition of $\nu(s_1, s_2)$ to any paths for which $\eta(s_1, s_2) > 0$. Let σ_1 and σ_2 be $\frac{1}{3}\eta$-approximations to s_1 and s_2 with general intersection, and σ'_1 and σ'_2 another such pair. Then $\sigma_i \simeq \sigma'_i$ in $R^2 - (s_j(0) \cup s_j(1))$, where $i = 1$ and $j = 2$ or *vice versa*. Hence $\nu(\sigma_1, \sigma_2) = \nu(\sigma_1, \sigma'_2)$ by Corollary 2, $= \nu(\sigma'_1, \sigma'_2)$ by Corollary 2 and property (C) of ν. The common value of these intersection numbers is defined to be $\nu(s_1, s_2)$. The reader will easily verify that the necessary approximations exist, and that properties (A), (B), (C), and Theorems 8·3, 8·4, 8·5 remain true when σ is replaced by s.

Theorem 8·7. *If l is a loop not passing through o, $l \simeq \mathbf{c}_k$ in $R^2 - (o)$ where $k = \omega(o, l)$.*

If $l(0) = e^{\alpha+i\beta}$, the deformation

$$d(\tau, \theta) = e^{-i\beta\theta} l(\tau)/(\theta| l(\tau) | + 1 - \theta)$$

in $R^2 - (o)$ carries l into a loop l' on the circle $| z | = 1$, having $l'(0) = 1$. We may therefore suppose l itself to be of this form.

Numbers $\tau_0, \tau_1, \ldots$ are now defined inductively as follows. First $\tau_0 = 0$. Supposing $\tau_{r-1} < 1$ defined so that $l(\tau_{r-1}) = \pm 1$ or $\pm i$, let τ_r be the least $\tau > \tau_{r-1}$ (if any) for which $l(\tau) = \pm il(\tau_{r-1})$; if there is none, $\tau_r = 1$. When $\tau_m = 1$ the series ends. Since $| l(\tau_r) - l(\tau_{r-1}) | = \sqrt{2}$ if $\tau_r < 1$, $\tau_r - \tau_{r-1}$ exceeds some fixed positive ϵ, and the series terminates with τ_m, for some $m < 1/\epsilon$. The arc $<\tau_{r-1}, \tau_r>$ of l covers a subset of the semicircular arc of $| z | = 1$ with mid-point $l(\tau_{r-1})$, and it is deformable on it, with fixed end-points, into the "quadrant path"

$$s_r(\tau) = l(\tau_{r-1}) e^{\pm\frac{1}{2}\pi i\tau};$$

and $l \simeq s_1 + s_2 + \ldots + s_m$ on $| z | = 1$. If both $+\frac{1}{2}\pi i\tau$ and $-\frac{1}{2}\pi i\tau$ occur as exponents there must be a consecutive pair with opposite exponents, $s_{r-1} + s_r = s - s$, which by 2·1 and 2·3 may be eliminated by a deformation on $| z | = 1$. Finally a loop is reached which is a sum of quadrant-paths all having the same exponent, and this loop is clearly c_k for some k. Since $l \simeq c_k$ in $R^2 - (o)$, $k = \omega(o, l)$.

Corollary 1. *If $\omega(o, l) = 0$, $l \simeq 0$ in $R^2 - (o)$.*

Corollary 2. *If l does not pass through o,*

$$\int_l \frac{dz}{z} = 2\pi i\omega(o, l).$$

This follows from 6·2 and the elementary special case $l = c_1$.

Corollary 2 may be regarded as the exact form of the statement that the ray oz turns through an angle $2k\pi$ when z describes a loop l which winds k times positively round o.

9. Theorem 9·1. *If x is a point of the inner domain of a simple loop l, $\omega(x, l) = \pm 1$.*

First suppose that l contains a straight segment: $l = s + ab$. Let x and y be points of the inner and outer domains of $| l |$ respectively, sufficiently near to $\frac{1}{2}(a + b)$ to ensure that the segment xy contains no point of s, and therefore crosses ab. Then by 8·5

$$\omega(x,l) - \omega(y,l) = \nu(xy,l) = \nu(xy,ab) = \pm 1.$$

Since y is in the outer domain of $|l|$, $\omega(y,l) = 0$, and therefore $\omega(x,l) = \pm 1$.

In the general case let c be a point of the inner domain, D, of $|l|$, and on any line through c let ab be the closure of the maximal interval in D containing c. Let s_1 and s_2 be the two arcs of l from a to b: $l = s_1 - s_2$. The cross-cut ab determines two domains, D_1 and D_2, in D, with frontiers $|s_1| \cup ab$ and $|s_2| \cup ab$. Let x be a point of D_1. Then

$$\omega(x,l) = \omega(x, s_1 - ab) - \omega(x, s_2 - ab) = \pm 1$$

by the previous case, since x is in the outer domain of $|s_2 - ab|$.

Corollary. *If l is a simple loop, l is not homotopic to $-l$ on $|l|$.*

It follows from Theorem 9·1 that simple loops may be classified as *positively or negatively oriented*, as the order of the inner domain is $+1$ or -1. The circular loop c_1, described in the "counter-clockwise" direction, is positively oriented according to this definition, in agreement with the usual convention.

Example. To evaluate the order of the inner domain of the triangular loop $ab + bc + ca$, a ray may be chosen issuing from a point inside the triangle near a, and passing near b. This ray meets bc and no other side. The condition for positive orientation is then that c is on the left of the ray ab, i.e. that it is on the same side of the line ab as $a + i(b - a)$. The condition for this is that $(c - a)/(b - a)$ have a positive imaginary part; or, if $a = \alpha_1 + i\alpha_2$, etc., that

$$\begin{vmatrix} \alpha_1 & \alpha_2 & 1 \\ \beta_1 & \beta_2 & 1 \\ \gamma_1 & \gamma_2 & 1 \end{vmatrix} > 0.$$

Thus the orientation of a triangle is fixed by choosing one of the two cyclical orderings of its vertices. This can be made the basis of an alternative treatment of the order of a point for a loop.

Theorem 9·2. *If a is a point of the inner domain of a simple loop l, l is not deformable to a point in $R^2 - (a)$.*

For if l_0 is a point path in $R^2 - (a)$, $\omega(a,l) = \pm 1$ but $\omega(a, l_0) = 0$.

Corollary. *If a, b are points in the two residual domains of a simple loop l in Z^2, l is not deformable to a point in $Z^2 - (a \cup b)$.*

Map Z^2 topologically on to itself so that b goes into the point at infinity.

Theorem 9·3. *If l_1 and l_2 are simple loops such that $|l_1| = |l_2|$, then $l_1 \simeq +l_2$ or $-l_2$ on $|l_1|$, as the loops have similar or opposite orientations.*

A first deformation on $|l_1|$ carries l_2 into a simple loop l' such that $l'(0) = l_1(0)$. The relation $l'(\tau') = l_1(\tau)$ between τ' and τ is a homeomorphism of the open interval $(0, 1)$ on to itself, which is extended to a homeomorphism ϕ of $< 0, 1 >$ by mapping the endpoints on themselves or on each other. Either $\phi(\tau)$ or $\phi(1-\tau)$ is linearly deformable into the identity on $< 0, 1 >$ with fixed end-points, say by $d(\theta, \tau)$; and $l(d(\theta, \tau))$ is a deformation of $\pm l'$ into l_1 on $|l_1|$.

If π is a simple polygon-chain on a grating in Z^2, lying in a domain D, the sides of π taken in order round the polygon determine a loop which may also be denoted by π.

Theorem 9·4. *A necessary and sufficient condition that $\pi \sim 0$ in D is that $\pi \simeq 0$ as a loop in D.*

Necessary. If K is the 2-chain bounded by π in D, $\mathscr{C}|K|$ is a simply connected domain, and there is a simple polygon, $|\pi_0|$, separating $|K|$ from $\mathscr{C}D$. One of the residual domains, D_0, of $|\pi_0|$ contains π and is contained in D. Since D_0 is simply connected, $\pi \simeq 0$ as a loop in D_0, (4·2), a fortiori in D.

Sufficient. If π (considered as a 1-cycle) does not bound in D, the two 2-chains bounded by π contain points a, b respectively of $\mathscr{C}D$, and by 9·2, Corollary, π is not null-homotopic in $Z^2 - (a \cup b)$, much less in D.

Corollary. *A necessary and sufficient condition for a domain D of R^2 or Z^2 to be simply connected is that every loop in D be null-homotopic in D.*

The following simple theorems are often useful. The three simple paths s_1, s_2, s_3 have common end-points, but do not meet otherwise.

Theorem 9·5. *If s_3 runs inside*† $|s_1 - s_2|$, *the loops $s_1 - s_2$ and $s_1 - s_3$ are similarly oriented.*

Let x_0 be any point of the inner domain of $|s_1 - s_3|$. Then x_0 is in the outer domain of $|s_2 - s_3|$, and

$$\omega(x_0, s_1 - s_2) = \omega(x_0, s_1 - s_3) + \omega(x_0, s_3 - s_2)$$
$$= \omega(x_0, s_1 - s_3).$$

Theorem 9·6. *A necessary and sufficient condition for s_3 to run inside $|s_1 - s_2|$ is that $s_1 - s_3$ and $s_2 - s_3$ have opposite orientations.*

Necessary. By 9·5 if s_3 runs inside $|s_1 - s_2|$, $s_1 - s_3$ and $s_2 - s_3$ have the orientations of $s_1 - s_2$ and $s_2 - s_1$.

Sufficient. If s_3 is outside $s_1 - s_2$ there is a point x_0 outside $s_1 - s_2$ and inside $s_1 - s_3$. Hence

$$0 = \omega(x_0, s_1 - s_2) = \omega(x_0, s_1 - s_3) - \omega(x_0, s_2 - s_3),$$

contrary to hypothesis.

10. *Complex integration (continued).* Let D be a domain in R^2 whose frontier is the union of the tracks of $k + 1$ disjoint, rectifiable, positively oriented simple loops $l_0, l_1, ..., l_k$, the outer boundary being $|l_0|$.

Theorem 10·1. *If $f(z)$ is regular in D and continuous in $\overline{D}$,*

$$\int_{l_0} f(z)\, dz = \sum_{j=1}^{k} \int_{l_j} f(z)\, dz.$$

Let D_j be the inner domain of $|l_j|$. It is easily seen that at least one of the inner loops, say l_1, can be joined to l_0 by two non-intersecting segments, $a_0 a_1$ and $b_0 b_1$, lying in D except for their end-points $(a_r, b_r \in |l_r|)$.‡ Let s_r, s_r' be the arcs $a_r b_r$ of l_r, $(l_r = s_r - s_r')$, and let $u = s_0 - (a_0 a_1 + s_1 + b_1 b_0)$, $u' = s_0' - (a_0 a_1 + s_1' + b_1 b_0)$. Since $s_0 - s_0'$ is positively oriented and $a_0 a_1 + s_1 + b_1 b_0$ runs in its inner domain, u is positively oriented, by 9·5; and similarly u' is negatively oriented.

The inner domains, D_u and $D_{u'}$, of u and u' meet neither D_1 nor each other.§ This follows from 9·5 and 9·6 as follows. (1) $s_0 - s_0'$ is positively

---

† I.e. $|s_3|$ is a cross-cut in the inner domain of $|s_1 - s_2|$.

‡ Let the shortest distance between $|l_0|$ and $\cup_1^k |l_r|$ be attained at c on l_c and c' on (say) l_1. If x and y are points, near c and c' respectively, of the outer domain of l_0, and of the inner of l_1, respectively, xy contains a segment $a_0 a_1$ as required, and any near parallel segment contains $b_0 b_1$.

§ I.e the arcs s_1 and s_1' are as shown in Fig. 79, and not interchanged. This result (the only point where essential use is made of the orientation of the l_r) is tacitly assumed in all proofs of 10·1 known to the author.

oriented, $v' = a_0a_1 + s_1' + b_1b_0$ runs inside $|s_0 - s_0'|$, therefore $s_0 - v'$ is positively oriented (9·5). Now

$$s_0 - v' \simeq (a_1a_0 + s_0 + b_0b_1) - s_1' \quad \text{on} \quad |s_0 - v'|.$$

Hence (2) $s_1 - s_1'$ and $(a_1a_0 + s_0 + b_0b_1) - s_1'$ are positively oriented, therefore s_1' runs outside $|s_1 - (a_1a_0 + s_0 + b_0b_1)| = |u|$, (9·6). Similarly s_1 runs outside $|u'|$. From this the stated results follow.

Thus if $l_{n_1}, \ldots, l_{n_h}$ are the loops l_r in D_u, $D_u - \cup \bar{D}_{n_i} \subseteq D$, and by an inductive hypothesis, starting with $k = 0$ (7·2)

$$\int_u f(z)\,dz = \sum_{r=1}^{h} \int_{l_{n_r}} f(z)\,dz;$$

and similarly, if $|l_{m_1}'|, \ldots, |l_{m_j}'| \subseteq D_{u'}$,

$$-\int_{u'} f(z)\,dz = \sum_{1}^{j} \int_{l_{m_r}'} f(z)\,dz,$$

since u' is negatively oriented. Therefore

$$\sum_{2}^{k} \int_{l_r} f(z)\,dz = \int_u - \int_{u'} = \int_{l_0} f(z)\,dz - \int_{l_1} f(z)\,dz.$$

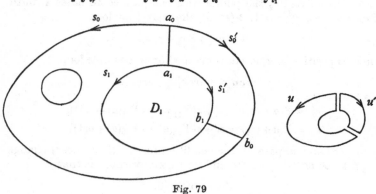

Fig. 79

11. Theorem 11·1. *If f is a topological mapping of R^2 on to an open subset of itself, there exists a fixed number $e = \pm 1$ such that $\omega(a, l) = e\omega(fa, fl)$ for all loops l and points a of $\mathscr{C}|l|$.*

Let dashed letters denote images, $f(a) = a'$, $f(l) = l'$, etc. The inner domain of $|\mathbf{c}_1|$ is mapped on to a simply connected domain which has $|\mathbf{c}_1'|$ as one component of its frontier, and must therefore be the inner domain of $|\mathbf{c}_1'|$. Hence $\omega(o', \mathbf{c}_1') = \pm 1 = e$, say. Clearly $\omega(o', \mathbf{c}_m') = me$.

If d is any deformation in R^2, fd is a deformation in the image set, and hence if $l_1 \simeq l_2$ in $R^2 - (a)$, $\omega(a', l_1') = \omega(a', l_2')$.

Let $\omega(a, l) = m$, and let $\mathbf{b}$ denote the circular loop $a + \alpha e^{2m\pi i\tau}$. By a trivial modification of 8·7, $l \simeq \mathbf{b}$ in $R^2 - (a)$. If α is large enough, o is an inner point of $\mathbf{b}$, and $\mathbf{b} \simeq \mathbf{c}_m$ in $R^2 - (o)$. It follows that

$$\omega(a', l') = \omega(a', \mathbf{b}') = \omega(o', \mathbf{b}') = \omega(o', \mathbf{c}_m') = me.$$

The number e is the *degree* of f. Simple loops have their orientations preserved or reversed as $e \gtrless 0$, and f is said to be sense-preserving or sense-reversing accordingly.

Theorem 11·2. *A necessary and sufficient condition for a continuously differentiable topological mapping*

$$Y(x): \quad \xi_1 + i\xi_2 \to Y_1(x) + iY_2(x)$$

of R^2 into itself to be sense-preserving is that the Jacobian be positive at at least one point.[28]

We use the notations of v. 22, and put x' for $Y(x)$. Let a first be any point, and e_1 and e_2 the points with complex coordinates 1 and i. If $c_r = a + \epsilon e_r$, for $r = 1, 2$ ($\epsilon > 0$), the triangular loop

$$l = ac_1 + c_1 c_2 + c_2 a,$$

which is positively oriented, is mapped on to a simple loop,

$$l' = \operatorname{arc} a'c_1' + \operatorname{arc} c_1'c_2' + \operatorname{arc} c_2'a'.$$

We have

$$c_1' = Y(a + \epsilon e_1) = a' + \epsilon(Y_{11}(a) + iY_{21}(a) + o(1)),$$
$$c_2' = Y(a + \epsilon e_2) = a' + \epsilon(Y_{12}(a) + iY_{22}(a) + o(1)).$$

Hence (Example, para. 9) the condition for the (rectilinear) triangle $a'c_1'c_2'$ to be non-degenerate and positively oriented is that

$$J(a) + o(1) > 0.$$

Since Y is a topological mapping, there is at least one point a at which $J(a) \neq 0$ (v. 22·6). If then ϵ is small enough the triangle $a'c_1'c_2'$ is non-degenerate, and the continuous differentiability ensures that a linear deformation of the arcs $a'c_1'$, $c_1'c_2'$, and $c_2'a'$ of l' into the straight paths with the same end-points does not pass over the centroid of the triangle; i.e. for such ϵ, l' and the triangle are similarly oriented.

Hence, first, if $J(a) > 0$, l' is positively oriented for small ϵ, and Y is sense-preserving. Secondly, if $J(a) \not> 0$, since $J(a) \neq 0$ l' is negatively oriented for small ϵ, and Y is sense-reversing.

NOTES

Comprehensive accounts of the theories to which this book is an introduction will be found in R. L. Wilder's *Topology of Manifolds* (1949) and G. T. Whyburn's *Analytic Topology* (1942), both Colloquium Publications of the American Mathematical Society, referred to as Wilder [1949] and Whyburn [1942] in the following notes. S. Lefschetz's *Introduction to Topology* (Princeton, 1949) provides a first introduction to the combinatorial theories used in Wilder's book.

1 (p. 2). Many writers use $\subset$ for "A is contained in B" (identity permitted). The symbol $\subseteq$, which we have preferred, has the advantage that in any context in which it is used there can hardly be any doubt which convention is being adopted. It also leaves the symbol $\subset$ available for proper subsets.

2 (p. 7). The calculus of sets can be expressed entirely in terms of equations, by defining "$A \subseteq B$" to mean $A = A \cap B$. Postulate B 5 is then to be replaced by $A = A \cap (A \cup B)$ and $A = A \cup (A \cap B)$; and it can then be inferred that $A \subseteq B$ is also equivalent to $B = A \cup B$. From this definition and B 1, 2, 4 and 5 (as modified), all of A and B 6·1 and 6·2 (but not B 3·1 and 3·2) are provable. A set of objects with operations $\cap$ and $\cup$ having these properties is a *lattice*. If in addition it has the distributive and complementation properties (B 3, C, D) it is a *Boolean Algebra*. In this form the calculus was discovered by George Boole (*The Mathematical Analysis of Logic*, 1847). See G. Birkhoff, *Lattice Theory*, 2nd edition, 1948.

3 (p. 11). (a) *A finite set is not similar to any proper subset of itself.* We may evidently suppose the finite set to be an I_n. The proof is by induction.

$n = 1$. The only proper subset of I_1 is the null-set, which is not similar to I_1.

$n > 1$. Suppose that (if possible) f is a (1, 1)-mapping of I_n on to the given proper subset, A. If $A \subseteq I_{n-1}$, f is a $(1, 1)$-mapping of I_{n-1} on to $A - f(n)$, a proper subset of A, and therefore a fortiori

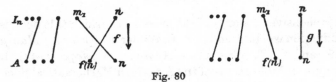

Fig. 80

of I_{n-1}. This is impossible, by the inductive hypothesis, and therefore A contains n. Let g be identical with f save that the correlates of n and $f^{-1}(n)$ are interchanged, i.e. if m_1 is the element of I_n correlated with n in A,

$$g(m_1) = f(n), \quad g(n) = f(m_1) = n,$$
$$g(r) = f(r) \text{ for all other values of } r.$$

Then g maps $I_n - (n)$, i.e. I_{n-1}, on to $A - (n)$, which, since $A \neq I_n$, is a proper subset of I_{n-1}. This again contradicts the inductive hypothesis.

It follows from this result that, if $m \neq n$, I_m is not similar to I_n. Thus *a non-null finite set is similar to one, and only one, of the sets* I_n. The corresponding n is called the *cardinal number*, or cardinal, of the set. The cardinal of the null-set is the number 0.

(b) *Every subset of a finite set is finite.* If the given set is not the null-set (whose only subset is itself) it may be taken to be a set I_n. The proof is by induction.

$n = 1$. The only subsets of I_1 are the null-set and itself, which are finite.

$n > 1$. If the given subset is contained in I_{n-1} it is finite by the inductive hypothesis. If not let it be $A \cup (n)$, where $A \subseteq I_{n-1}$. By hypothesis there is a $(1, 1)$-mapping of A on to a set I_r, and by mapping n on $r + 1$ we obtain a mapping of $A \cup (n)$ on to I_{r+1}.

(c) *If A and B are finite $A \cup B$ is finite.* If A and B do not meet we have only to add the cardinal number of A to the suffix of each element of B (in some enumeration) to obtain a $(1, 1)$-mapping of $A \cup B$ on to an I_n. In the general case

$$A \cup B = A \cup (B - A), \quad A \cap (B - A) = 0,$$

and $B - A$ is finite by (b).

It follows by induction on the number of sets that *the union of any finite set of finite sets is finite.*

4 (p. 16). The theory in § 2 is a fragment of the theory of infinite cardinal numbers, which contains a highly developed arithmetic, with inequalities, addition, multiplication and exponentiation. See, e.g., Sierpiński, *Nombres Transfinis* (1928).

5 (p. 19). The axioms to be satisfied by the operations $+$ and $\cdot$ (the latter sign usually omitted) are

I 1. $(x+y)+z = x+(y+z)$, $\quad x+y = y+x$;
  2. there exists an o such that $x+o = x$ for every x;
  3. given any x, there exists a y such that $x+y = o$.

II 1. $\lambda(\mu x) = (\lambda\mu)x$;
  2. $(\lambda+\mu)x = \lambda x+\mu x$, $\quad \lambda(x+y) = \lambda x+\lambda y$;
  3. $1x = x$.

If, for any x and q, $x+q = x$, then $q = o$. For let $y+x = o$ (I 3). Then $q = (y+x)+q = y+(x+q) = y+x = o$.

The element y corresponding to a given x in I 3 *is unique.* For if also $x+z = o$, then $y = y+(x+z) = (x+y)+z = z$.

For any x, $x+(0 \cdot x) = (1+0)x = 1x = x$. Hence $0x = o$, and if $x = b+(-1)a$, then $a+x = b+(1-1)a = b+0a = b$.

For any $\lambda \neq 0$, $\lambda o+x = \lambda(o+\lambda^{-1}x) = \lambda(\lambda^{-1}x) = x$. Hence $\lambda o = o$.

For the theory of vector spaces see Banach, *Théorie des opérations linéaires* (1932).

6 (p. 26). (*a*) Let $b_1, b_2, ..., b_k$ be linearly independent points of R^p, and let the coordinates of b_r be β_{ir}. The columns of the matrix $\{\beta_{ir}\}$ are linearly independent. Since the matrix has only p rows it cannot have more than p linearly independent columns, i.e. $k \leqslant p$. Thus the algebraic dimension cannot exceed p. On the other hand, the points $e_1, e_2, ..., e_p$, where e_j has its jth coordinate 1 and the rest 0, are clearly linearly independent.

(*b*) Let E^k be the set of points

$$(1) \qquad \xi_i = \alpha_i + \sum_{r=1}^{k} \tau_r \beta_{ir} \quad (i = 1, 2, ..., p),$$

where the α_i and β_{ir} are fixed (the β_{ir} forming k linearly independent columns of p numbers), and the τ_r take all real values. Then
there exists a non-singular square matrix B of order p, whose first
k columns are the β_{ir} $(i = 1, 2, ..., p; r = 1, 2, ..., k)$. Let the
elements of B still be called β_{ir} (where now $r = 1, 2, ..., p$), and let
$\check{\beta}_{ir}$ be the elements of the inverse matrix. Then

$$(2) \qquad \sum_1^p \check{\beta}_{hi}(\xi_i - \alpha_i) = \sum_1^k \delta_{hr}\tau_r$$

$$= \begin{cases} \tau_h \text{ if } h \leqslant k, \\ 0 \text{ if } h = k+1, ..., p. \end{cases}$$

Thus the coordinates of all points of E^k satisfy the $p - k$ equations

$$(3) \qquad \sum_1^p \check{\beta}_{hi}(\xi_i - \alpha_i) = 0 \quad (h = k+1, ..., p),$$

which are linearly independent, since the $\check{\beta}_{hi}$ involved form
columns of a non-singular square matrix. Conversely if x is any
point whose coordinates satisfy (3), and τ_h is defined to be
$\Sigma \check{\beta}_{hi}(\xi_i - \alpha_i)$ for $h = 1, 2, ..., k$, the coordinates of x satisfy the
equations (2), a system with non-vanishing determinant whose
solution is (1). Therefore $x \in E^k$.

In the particular case of the straight line joining a to $a + b_1$ the
point ξ_i has (with the above notations) the parameter

$$\tau_1 = \sum_{i=1}^p \check{\beta}_{1i}(\xi_i - \alpha_i).$$

Therefore the ray $\tau_1 \geqslant 0$ is the set of points determined by the
$p - 1$ equations of the line and the inequality

$$\sum_{i=1}^p \check{\beta}_{1i}(\xi_i - \alpha_i) \geqslant 0.$$

A segment, being the common part of the rays $\tau_1 \geqslant 0$ and $\tau_1 \leqslant 1$,
is determined by $p - 1$ equations and two inequalities between
the ξ_i.

(c) That E^k is a linear subset is obvious, for if x and y correspond to the values σ_r and τ_r of the parameters, $x + \lambda(y - x)$
corresponds to the parameters

$$\sigma_r + \lambda(\tau_r - \sigma_r).$$

Suppose now that E is a linear subset of R^p, i.e. a set satisfying the "straight line" condition that if it contains x and y it contains the line xy. Let a be any point of E, and $b_1, b_2, ..., b_k$ a maximal set of linearly independent points of R^p such that all the points $a + b_r$ ($r = 1, 2, ..., k$), belong to E. Such a set can be found by taking b_1 to be any point, other than o, such that $a + b_1 \in E$, and adding one by one points with the property, linearly independent of those already chosen, until no more exist. This gives a maximal set in at most p steps.

The points $$a + \sum_1^k \tau_r b_r$$

belong to E for all real values of the τ_r. By the "straight line" condition this is true when $k = 1$, and since

$$a + \sum_1^k \tau_r b_r = \frac{1}{2}\left(a + \sum_1^{k-1} 2\tau_r b_r\right) + \tfrac{1}{2}(a + 2\tau_k b_k),$$

it is on a line joining two points which, by an inductive hypothesis, belong to E.

The points $$a + \sum_1^k \tau_r b_r$$

include all points of E. For if x is any point of E the points

$$x - a, \ b_1, \ b_2, \ ..., \ b_k$$

are all of them points y such that $a + y \in E$. Therefore, since the b's form a maximal linearly independent set with this property,

$$\lambda_0(x - a) + \sum_1^k \lambda_r b_r = o,$$

where the λ's are not all zero; and $\lambda_0 \neq 0$, since the b_r's are linearly independent. Hence division by λ_0 is permissible, and

$$x = a + \sum_1^k \tau_r b_r,$$

where $\tau_r = -\lambda_r / \lambda_0$.

7 (p. 32). The condition established in Theorem 9·1 can be used to define the derived set (and was so used in the first edition of this book). In certain "bad" topological spaces (see Note 10) the two definitions are not equivalent, but in such spaces the derived set is not a useful notion.

8 (p. 33). See Lebesgue, *Leçons sur l'intégration* (1928), p. 315.

9 (p. 40). This definition of compactness is not suitable for general topological spaces. See Note 10.

10 (p. 59). If "open set" is taken as the fundamental undefined notion, the name *topological space* is given to sets in which certain subsets have been selected as "open sets", subject to the following conditions:

(1) the union of any set of open sets is open,
(2) the intersection of two open sets is open,
(3) the empty set and the whole space are open sets.

The notions "closed set" and "closure of a set" may equally well be taken as primitive, and lead, with suitable definitions and axioms, to exactly the same class of spaces. To obtain objects resembling the spaces of ordinary geometry and analysis a *separation axiom* must be added, e.g. Hausdorff's axiom: *any two distinct points are contained in disjoint open sets.*

If "convergent sequence" is taken as primitive notion, a different class of spaces is obtained, the *convergence spaces*, which include metrisable spaces but otherwise diverge in their properties from the class of topological spaces.

For the purposes of general topology the "topological spaces" have been found the more suitable, and the definitions in this book are for the most part expressed in a form that can be used without modification in general topological spaces. An exception is the definition of *compactness*, now usually defined, for general spaces, to mean that the Heine-Borel-Lebesgue covering theorem holds in the space. As Theorem II. 15·3 shews, this definition is equivalent to ours in metrisable spaces, but it fails to be so in general spaces. The name "sequentially compact" is used in such cases for the property defined in the text.

The fact that most of our theorems remain significant in general topological spaces does not, of course, ensure that they remain true.

For an account of the "point-set" properties of general spaces see Alexandroff and Hopf, *Topologie I* (1935); Kuratowski, *Topologie I* (2nd ed. 1949), including convergence spaces;

Bourbaki, *Eléments de mathématique* (in progress), livre III (Topologie générale), with a generalisation of convergence of sequences suitable for non-separable spaces.

11 (p. 81). A proof was given in the first edition of this book, IV. 7·3.

12 (p. 84). The properties of locally connected plane sets have been extensively studied by R. L. Moore (see his *Foundations of Point Set Theory*, Amer. Math. Soc. Colloquium Publ. 1932) and G. T. Whyburn [1942].

13 (p. 92). See Whyburn [1942] II. 5·1, and references there given.

14 (p. 93). Janiszewski, *Journal de l'Ecole Polyt.* 16 (1912), 76–170. For theorems related to the contents of this section, in particular on the relation of a set to its cut-points, see Whyburn [1942], Chapter III.

15 (p. 99). A. J. Ward, *Proc. Lond. Math. Soc.* 41 (1936), 191.

16 (p. 110). J. W. Alexander, *Trans. Amer. Math. Soc.* 23 (1922), 342.

17 (p. 115). C. Jordan was the first to state, and to attempt to prove, the theorem that bears his name. The first correct proof was given by Veblen in 1905. The text proof, which is that of Alexander, is based on Brouwer's proof (*Math. Annalen*, 1909).

18 (p. 120). Exercise 6 is (for $p = 3$) Kuratowski's Theorem on three continua, *Monatshefte für Math.* 30 (1929), 77. This and other theorems of Ch. V were extended by Eilenberg (*Fundam. Math.* 26 (1936), 61) to sets which are neither open nor closed.

19 (p. 135). A theorem resembling 21·2 was stated, with an outline of a proof, by Lebesgue in 1911 (*C.R. Acad. Sci., Paris*, 152, p. 841). It is the special case for a j-sphere of the Alexander Duality Theorem (loc. cit. Note 16).

20 (pp. 138 and 139). Alexandroff and Hopf, *Topologie I*, pp. 477–8, give a proof of the existence and (substantially) of the uniqueness of a solution of $Y(x) = y$ under the conditions of 22·1, using methods connected with the *order* of a point for a cycle. *Proof of* 22·3. If x' and x are near enough to a, and z to c, it follows from (1) of 22 that

$$\xi'_j - \xi_j = \sum_{i=1}^{p} (\eta'_i - \eta_i)\, \breve{Y}_{ji}(u_i),$$

where $\{\breve{Y}_{ij}\}$ is the inverse matrix of $\{Y_{ij}\}$, and u_i is on the segment from (x, z) to (x', z). From the continuity of X as a function of y and z it follows that as $y' \to y$, $\breve{Y}_{ji}(u_i) \to \breve{Y}_{ji}(x)$. Therefore X_j is differentiable and

$$\frac{\partial X_j}{\partial \eta_i} = \breve{Y}_{ji}(x),$$

which is a rational function of the Y_{ij} with non-vanishing denominator, and therefore continuous.

Proof of 22·4. For x, x' near a, and z, z' near c, we have

$$Y_i(x', z') - Y_i(x, z) = \sum_{j=1}^{p} (\xi'_j - \xi_j)\, Y_{ij}(u_i) + \sum_{h=1}^{q} (\zeta'_h - \zeta_h)\frac{\partial Y_i}{\partial \zeta_h}(u_i),$$

where u_i is on the segment (x, z) to (x', z'). Hence

$$\xi'_j - \xi_j = \sum_{j=1}^{p} (\eta'_i - \eta_i)\, \breve{Y}_{ji}(u_i) - \sum_{i=1}^{p}\sum_{h=1}^{q} (\zeta'_h - \zeta_h)\frac{\partial Y_i}{\partial \zeta_h}\breve{Y}_{ji},$$

from which the result follows.

Proof of 22·5. We assume that

$$\frac{\partial(Y_1, Y_2, \ldots, Y_r)}{\partial(\xi_1, \xi_2, \ldots, \xi_r)} \neq 0$$

in $U(a)$. For any x let x^1 and x^2 denote the points $(\xi_1, \ldots, \xi_r)$ and $(\xi_{r+1}, \ldots, \xi_p)$ of R^r and R^{p-r} respectively, so that $x = (x^1, x^2)$. The first r equations $Y(x) = y$ can then be written $Y^1(x^1, x^2) = y^1$. Considered as equations for x^1 they have, by 22·2, the unique solution $x^1 = X^1(y^1, x^2)$, if y^1 and x^2 are sufficiently near to b^1 and a^2. Consider the function

$$W(y^1, x^2) = Y^2(X^1(y^1, x^2), x^2).$$

W is a function of y^1 alone. (Throughout what follows we let

$$i, j, k = 1, \ldots, r; \quad s, t = r+1, \ldots, p; \quad m, n = 1, \ldots, p.)$$

$$\frac{\partial W_s}{\partial \xi_t} = \sum_{i=1}^{r} \frac{\partial Y_s}{\partial \xi_i}\frac{\partial X_i^1}{\partial \xi_t} + \frac{\partial Y_s}{\partial \xi_t}$$

$$= - \sum_{i,j=1}^{r} Y_{si}\, \breve{Y}_{ij}\, Y_{jt} + Y_{st}$$

NOTES   207

by 22·4 where $\{\breve{Y}_{ij}\}$ is the inverse of the r-rowed non-singular matrix $\{Y_{ij}\}$. Since the full matrix $\{Y_{mn}\}$ is of rank r, and its first r rows are linearly independent, there exist functions f_{si} such that

$$Y_{s,\,n} = \sum_{i=1}^{r} f_{si} Y_{in}.$$

Hence
$$\frac{\partial W_s}{\partial \xi_l} = - \sum_{i,\,j,\,k=1}^{r} f_{sk} Y_{ki}\, \breve{Y}_{ij} Y_{jl} + Y_{sl}$$

$$= - \sum_{j,\,k=1}^{r} f_{sk} \delta_{kj} Y_{jl} + \sum_{i=1}^{r} f_{si} Y_{il} = 0.$$

Thus $W(y^1, x^2)$ is in fact $W(y^1)$, and

$$W_s(Y_1(x), ..., Y_r(x)) = Y_s(X^1(Y^1, x^2), x^2) = Y_s(x),$$

from the uniqueness of X^1 if x is near a. Finally

$$\frac{\partial W_s}{\partial \eta_i} = \sum_{j=1}^{r} \frac{\partial Y_s}{\partial \xi_j} \frac{\partial X_j^1}{\partial \eta_i} = \sum_{j=1}^{r} Y_{sj}\, \breve{Y}_{ji}.$$

These are continuous functions of x, i.e. (putting X^1 for x^1) of (y^1, x^2), and hence, by what has been proved, of y^1.

21 (p. 152). The properties of groups required in the text are established in all texts on group theory, but the proofs for the special class of groups involved are so simple that they are given here. The reader is assumed familiar with the notion of abelian groups, and of generators and relations of a group. The group operation is denoted by ' $+$ '.

We consider an abelian group $\mathfrak{G}$ in which every element g satisfies $g + g = 0$. The equations $g_1 + g_2 = 0$ and $g_1 = g_2$ are therefore equivalent, and for any integers m_i and elements g_i,

$$\sum_{1}^{k} m_i g_i = \sum_{1}^{k} r_i g_i,$$

where r_i = residue $m_i \bmod 2$. Hence if $\mathfrak{G}$ has a finite set of generators it is finite. The elements $g_1, ..., g_k$ are *linearly dependent* if there exist integers r_i, not all even, such that $\sum_{1}^{k} r_i g_i = 0$.

I. *If $\mathfrak{G}$ has a finite set of generators* $a_1, a_2, ..., a_q$, *every set of $q + 1$ elements of $\mathfrak{G}$ is linearly dependent*. Let $g_1, g_2, ..., g_{q+1}$ be any elements, and let

$$g_i = \sum_{1}^{q} n_{ir}\, a_r.$$

The set of q (numerical) equations

$$\sum_{i=1}^{q+1} x_i n_{ir} = 0$$

in the $q+1$ variables x_i has at least one solution $x_i = h_i$, where the h_i are rational and not all zero. By multiplication by a suitable common factor the h_i may be made integers, not all even; and

$$\sum_i h_i g_i = \sum_{i,r} h_i n_{ir} a_r = 0.$$

II. *If every set of $q+1$ elements in $\mathfrak{G}$ is linearly dependent the group has a finite basis (linearly independent set of generators)* a_1, a_2, ..., a_m, *where* $m \leqslant q$. If 0 is the only element the theorem is trivial. If not, let $a_1 \neq 0$, and let a_2, a_3, ... be successively chosen, each linearly independent of its predecessors, until no further such elements exist. This must happen for some $m \leqslant q$. If g is any element of $\mathfrak{G}$

$$ng + \sum_1^m x_i a_i = 0$$

for some n and x_i, not all even. If n is even, $ng = 0$, i.e. $\sum_1 x_i a_i = 0$ for x_i not all even, contrary to hypothesis. Therefore n is odd, i.e.

$$g = \sum_1^m x_i a_i.$$

Thus the (a_i) are a set of generators.

From I and II it follows that either $\mathfrak{G}$ is infinite, there is no finite set of generators, and (by II) there exist arbitrarily large linearly independent sets; or $\mathfrak{G}$ is finite, there exists at least one finite basis, and a maximum possible size, p, for linearly independent sets. The number p is the *rank* (mod 2) of $\mathfrak{G}$. It has been shewn in I and II that *every basis contains exactly p members.* From I it follows that p *is the minimum possible size for a set of generators of* $\mathfrak{G}$.

22 (p. 159). For a fuller account of the properties of plane sets touched on in this section see Whyburn [1942], Chap. VI. The notion of local connection was extended to higher dimensions by Lefschetz (1930). "S is q-LC at a" means that every mapping of a q-sphere into a sufficiently small neighbourhood of a can be shrunk to a point within an assigned neighbourhood of a. If a

"0-sphere" is understood to be a pair of points, "locally connected" and 0-LC are the same in locally compact spaces (end of IV. 9). The capital letters LC distinguish this "homotopy" property from the corresponding "homology", or bounding, property, q-lc. These ideas have led to generalisations, particularly by R. L. Wilder (see Wilder [1949]), of some of the theorems of this section to sets in R^p.

23 (p. 166). A proof was given in the first edition of this book, VII. 3·5, Case II.

24 (p. 166). Schoenflies (*Gött. Nachr., Math. Phys. Kl.*, 1902, p. 185). A proof similar to that in the text by Hilbert, *Math. Annalen*, 56 (1903), 381.

25 (p. 168). Kerékjártó, *Topologie I* (1923), p. 87.

26 (p. 186). M. H. A. Newman, Path-length and linear measure, *Proc. Lond. Math. Soc.* (3) 2 (1952), 455–68.

27 (p. 187). First proved under these conditions by S. Pollard, *Proc. Lond. Math. Soc.* 21 (1923), 456–82; the corresponding form of Green's Theorem by W. Gross, *Monatshefte für Math.* 27 (1916), 70–120.

28 (p. 198). Cf. Alexandroff and Hopf, loc. cit. Note 20.

INDEX

Symbols with a fixed meaning